スッキリ！がってん！電気英語の本

春日　健・濱﨑　真一 [著]

電気書院

はじめに

　"電気があれば何でもできる"とまではいわないが，ある日突然電気が使えなくなったら何が起こるだろうか．身近なところでは，コンピュータやテレビから情報は得られず，伝書鳩に頼らざるを得ないかもしれない．日常全般の暮らしは昔のやり方に戻るしかないだろうし，例えば冷蔵庫はただの箱と化すのは間違いない．そして，工場では機械やロボットはただの物体と化し，病院では医療機器などの使用不能により致命的な結果となることは目に見えている．今日，電気に依存している我々にとっては死活問題である．この本は，電気の学習を通じて英語の学習もでき，うまくいけば，その両方に精通できるのではないかと目論んでいる．

　電気英語の核は3Cの考え方，すなわち Clear，Concise，Correct といわれている．電気英語では，誤解を与えない短い表現が特に重要である．

　読者のなかには，外国製品を購入したところ，英文のマニュアルだけが付属していたという経験をされた方もいるのではないだろうか．外国製品を使いこなすにもマニュアルを読んで理解する英語の読解力が必要となる．一方，日本製品を輸出する場合にも英文マニュアルがないと買ってもらえないのも当然である．また，海外の技術研究分野の文献を読む必要のある方，国際会議で研究発表を予定されている方や海外の学会に論文を投稿予定の方にとっても英語での発表や執筆は必須である．さらには，インターネット全盛のこの時代，外国人との電子メールのやり取りにも英語でなければ話が通じ

ない．このように，英語の読解力と英作文力はますます必要となってきている．

　本書では，このような新しい技術の基礎である電気の英語表現をわかりやすく取りあげている．電気英語に興味をもたれる方や電気英語をこれから学びたい方を対象に，身近な性質，現象などの電気の一般知識を英語と日本語で理解できるように解説している．高校程度の学力があれば理解できるように，できるだけわかりやすく解説し，専門書を読み解くための体系的で確実な基礎知識を届けることを目指している．

　第1章では，電気の基礎，第2章では，直流回路と回路素子，第3章では，電気回路における物理現象，第4章では交流回路を取りあげている．

2021年10月　著者記す

目　次

3 電気回路における物理現象

電気との基礎

この章では，はじめに電気とは何かということで，静電気，原子と電子について取りあげる．次に，電気を取り扱うのに必要な電流，電圧，受動素子である抵抗器，コンデンサ，コイルを取りあげる．そして，電気エネルギーと熱エネルギーの関係を表すジュールの法則と，史上最初の電池であるボルタ電池といろいろな電池を紹介する．

1.1　静電気

空気が乾燥している冬，じゅうたんを敷いた廊下を歩いて，ドアのノブに手を触れた瞬間にビリッとくることがある．これは物質に電気が帯電したもので静電気と呼ばれる．

異なった種類の物質の間で摩擦をすると，物質中の電子がエネルギーを受けて運動が活発になり，一方の物質へ移動する．この結果，一方の物質に正の電荷，もう一方に負の電荷が帯電して紙が吸い付く．このことは，異種の静電気が発生して，お互いに引き合うために起こる現象である．

The electricity charged on an object is called electric charge.

● 物体に帯電した電気は，電荷と呼ばれる．

There are two types of electric charges : positive and negative.

● 電荷には正と負の2種類がある．

Electric charges of the same type repel each other, and electric charges of different types attract each other. The closer the distance, the greater the force.

● 同種の電荷は互いに反発し，異種の電荷は互いに引き付け合う．距離が
近いほど力は大きくなる．

First, touch the antistatic sheet.

● はじめに帯電防止シートに触れなさい．

　二つのコルク球を糸でつるし，絶縁体であるガラス棒を絹布でこ
すって両方のコルク球に近づけると反発する．また，一方のコルク
球にガラス棒を，もう一方のコルク球に絹布を近づけると引き合う．
真空中または空気中で，二つの点電荷をそれぞれ Q_1，Q_2 [C] とし，
電荷間の距離を r [m] とすると，これらの間に働く静電力 F [N] は次
のように表される．これをクーロンの法則（Coulomb's law）という．

$$F = \frac{1}{4\pi\varepsilon} \cdot \frac{Q_1 Q_2}{r^2} \text{ [N]}$$

　ここで，ε は誘電率と呼ばれ，電荷が存在する空間を占める媒質に
よって決まる定数である．この式で表される力 F をクーロン力（静
電力，静電気力）という．

Coulomb's law states that the magnitude of the electrostatic force
between two point charges is directly proportional to the product
of the magnitudes of the charges and inversely proportional to the
square of the distance between them.

● クーロンの法則は二つの点電荷間の静電力がそれらの電荷の積に比例し，
電荷間の距離の2乗に反比例することを表している．

1.2 原子と電子

　すべての物質は，約100種類の原子により構成されている．その
原子は，種類により異なるが，約 10^{-9} m の大きさからなり，中心

には正電荷をもった一つの原子核が，その周囲には負の電荷をもったいくつかの電子が円軌道を描くモデルで考えられている．

All physical objects are made up of atoms. Inside an atom are protons, electrons and neutrons. The protons are positively charged, the electrons are negatively charged, and the neutrons are neutral.

Every atom is composed of a heavy atomic nucleus and one or more light electrons that orbit around it. The atomic nucleus is composed of one or more neutrons and one or more protons.

The number of protons in the nucleus is dependent on the type of element. The atomic number equals the number of protons in the nucleus.

● 　あらゆる物は原子でできている．原子の内部には，陽子，電子，そして中性子がある．陽子は正に，電子は負にそれぞれ帯電し，中性子は電気的な性質がなく中性である．
　すべての原子は，重い原子核とその周りを回る一つ以上の軽い電子で構成される．原子核は，一つ以上の中性子と一つ以上の陽子から構成される．
　原子核の中の陽子の数は，元素の種類によってすべて異なる．原子番号は，原子核にある陽子の数に等しい．

As an example of the simplest atom, Fig.1・1 shows the atomic structure of a hydrogen atom. The hydrogen atom is composed of one electron and one atomic nucleus consisting of one proton. The

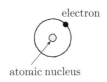

図1・1　Atomic structure of a hydrogen atom.

electron is shown in an orbital ring around the nucleus. In order to account for the electrical stability of the atom, we can consider the electron as spinning around the nucleus, as the planets revolve around the sun.

● 　最も単純な原子の例として，図1・1に水素原子の原子構造を示す．水素原子は，1個の電子と1個の陽子から成る1個の原子核で構成される．電子は，原子核の周りの軌道環上に示される．原子の電気的安定性を説明する上で，惑星が太陽の周りを回るように，原子核の周りを回る電子を考えることができる．

　また，シリコンの原子核は，14個の陽子と14個の中性子で構成され，その周囲を14個の電子が回転している．この原子の原子構造を図1・2に示す．原子核の周りの電子は，いくつかの層に分かれて存在し，これらの層を電子殻という．

電子

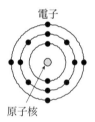

原子核

図1・2　シリコン原子の原子構造

Valence electrons are present in the outermost shell of an atom. They are more weakly attracted to the positive atomic nucleus than are the inner electrons, and hence tend to be free electrons by given little energy from the outside.

● 　価電子は，原子の最外殻にある．それらは内側の電子と比較して正の原子核との結びつきが弱い．したがって，外部からのわずかなエネルギーによって自由電子になりやすい．

1.3 電流

すべての物質は原子で構成され，原子の中心には原子核があり，その周囲の軌道を電子が回っている．例えば，ナトリウム原子の最外殻の価電子の数は1で，その結果，価電子に対する原子核の結合力は弱く，外部からエネルギーが加わると，価電子は軌道を外れて金属結晶の間を自由に動き回るようになる．このような電子は自由電子と呼ばれ，一般に，金属原子は自由電子を発生しやすい傾向にある．

電流は電子の流れということだが，電子は電荷をもっているので，電流は電荷の移動ということができる．

Electric current is supposed to flow from positive terminal to negative terminal, but in fact electrons flow from negative to positive.

● 電流は正極から負極へ流れることになっているが，実際は電子が負極から正極へ流れる．

When electrons can move easily from atom to atom in a material, it is called a conductor. Copper is an extremely better conductor. On the other hand, an insulator is a material that restricts the flow of electrons. Some examples of insulators are glass and most rubbers.

● 電子が物質内の原子から原子へ自由に移動できると，その物質は導体と呼ばれる．銅は良導体である．一方，絶縁体は，電子の流れを妨げる物質である．絶縁体の例を挙げると，ガラスやほとんどのゴムがある．

図1・3に示すように，銅線に電源を接続すると自由電子は電源の正極に引き付けられ，負極からは電子が連続的に供給される．この電子の流れが電流である．すなわち，電源の正極は電子を受け入れる側，負極は電子を供給する側である．

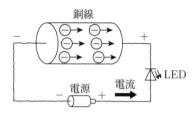

図1・3　金属内の電子の移動

The International System of Units(SI) is a scientific method of expressing the magnitudes or quantities of important natural phenomena. There are seven base units in the system. This system was made on the basis of meter-kilogram-second-ampere (MKSA).

Electric current is the rate of flow of electric charges through a conductor. The SI unit of electric current is ampere. One ampere is equal to the flow of one coulomb of electric charge per second.

● 　国際単位系（SI）は，重要な自然現象に関する大きさや量を表現する科学的方法である．この単位系には，七つの基本単位がある．この単位系は，メートル・キログラム・秒・アンペア（MKSA）を基礎として構成された．

　　電流は，導線を通過する電荷の移動率である．国際単位系では，電流はアンペアで表される．1アンペアは1秒間に1クーロンの電荷の移動量に等しい．

1.4　電圧

　電流はプラスからマイナスに流れると定義され，この電流を流す電気的な圧力を電圧という．

Electric potential difference is another word for voltage. The potential difference of a power supply is an electrical pressure that causes a current to flow in a circuit.

● 　電位差は，電圧の別の表現である．電源の電位差とは，回路で電流を発生させる電気的な圧力のことである．

　電気回路の例を図1・4に示す．電流は，矢印で示したように，電位の高い方から低い方へ流れる．また，電位差や起電力は，電位の高い方に矢印を向けて表す．

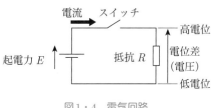

図1・4　電気回路

　Electromotive force is defined as the electric pressure that makes a current flow in a circuit.

● 　起電力は，回路に電流を流す電気的な圧力と定義される．

　How should an ammeter and a voltmeter be connected in a circuit?

● 　回路で，電流計や電圧計はどのように接続すればよいか．

　電流計や電圧計には，＋端子と−端子があり，電流が＋端子から内部を通って−端子へ流れるように，図1・5に示すように接続される．

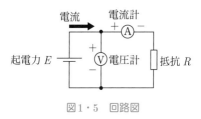

図1・5　回路図

The ammeter is connected in series with the circuit, so that the current to be measured flows directly through the ammeter. You have to understand that an ammeter has a very low resistance to minimize the voltage drop through the ammeter. If the ammeter is connected in parallel with the circuit, damage to the ammeter results from extremely high current.

● 電流計は，回路に直列に接続される．その結果，測定される電流は，そのまま電流計を流れる．電流計での電圧降下を最小にするために，電流計は非常に小さい抵抗をもっていることを理解しなければならない．もし，電流計が回路に並列に接続されると，過大電流によって電流計は破損する．

A voltmeter is a tool used to measure the potential difference between two points in a circuit. The voltmeter is connected in parallel with the element to be measured. You have to understand that a voltmeter has a very high resistance so as to minimize the current flow through the voltmeter.

● 電圧計は，回路内の2点間の電位差を測定するために用いられる．電圧計は，測ろうとする素子と並列に接続される．電圧計を流れる電流を最小限に抑えるため，電圧計は非常に高い抵抗を有することを理解しなければならない．

1.5 抵抗器

抵抗器は，回路に流れる電流を一定に保ったり，必要に応じて変化させたりするための電子部品である．

Resistors are the most commonly used elements in electric and electronic circuits. These are electronic components that limit the flow of electric current and divide the voltage. Moreover, resistors are called passive elements because they do not require a source of energy to perform their purposeful functions.

Resistors can be mainly classified based on the type of material used, the rated power and resistance value. The rated power indicates how much power a resistor can handle before it becomes too hot and burns up. Power is measured in units called watts.

The tolerance of a resistor is the maximum difference between its actual value and the required value and is generally expressed as a plus or minus percentage value.

● 抵抗器（抵抗）は電気回路や電子回路で最も普通に用いられる素子である．これらは電流を制限し，電圧を分割する電子部品である．さらに，抵抗器は目的とする機能を実現するためにエネルギー源を必要としないため，受動素子と呼ばれる．

　抵抗器は主に使用される材料のタイプ，定格電力および抵抗値に基づいて分類される．定格電力は，抵抗器が過熱して焼損しない最大の電力を表す．電力は，ワットと呼ばれる単位で測定される．

　抵抗値許容差は，実際の値と必要な値との最大の差であり，一般にプラスまたはマイナスのパーセンテージで表される．

Resistors are mainly classified into fixed resistors and variable resistors. Fixed resistors are resistors with fixed resistance value that cannot be changed. On the other hand, variable resistors are resistors that the resistance value can be adjusted.

● 抵抗器は，主に固定抵抗器と可変抵抗器に分類される．固定抵抗器は抵抗値を変更することのできない，固定された抵抗値をもつ抵抗器である．一方，可変抵抗器は，抵抗値を調整することができる抵抗器である．

In electronic circuits, many different types of resistors have been used. Some of these are described below.

Carbon film resistors are resistors are by depositing a thin layer of carbon on an insulated substrate.

Metal film resistors are made by depositing metal film on the ceramic substrates. Metal film resistors have less variation in resistance value and have higher stability.

Metal oxide film resistors use metal oxide as the resistive material within the resistors. The metal oxide film resistors can withstand higher temperature than carbon or metal film resistors.

Wire wound resistors are made by winding high-resistance wire around a cylinder of insulated material. They can provide very high power ratings and can operate at very high temperatures.

Surface mount resistors, also called chip resistors, are very small rectangular-shaped metal oxide film resistors designed to be soldered directly onto the surface of circuit board.

Fusible resistors are wire wound resistors that are designed to burn open easily when the rated power is exceeded. They fulfill a dual function as both fuse and resistor.

● 電子回路では，多くの異なるタイプの抵抗器（抵抗）が使用されている．これらのうちのいくつかを以下に説明する．

炭素皮膜抵抗器（カーボン抵抗）は，セラミック基板上に炭素皮膜を蒸着させた抵抗器である．

金属皮膜抵抗器は，セラミック基板上に金属皮膜を蒸着させることによって製造される．金属皮膜抵抗器は，抵抗値のばらつきが小さく，安定性が高い．

酸化金属皮膜抵抗器は，抵抗器内の抵抗材料として金属酸化物を使用する．酸化金属皮膜抵抗器は，炭素または金属皮膜抵抗器よりも高い温度に耐えることができる．

巻線抵抗器は，高抵抗線を絶縁材料でできた円筒の周りに巻くことによってつくられる．この抵抗器は，非常に高い電力が得られ，かつ非常に高い温度で動作することができる．

チップ抵抗器とも呼ばれる表面実装抵抗器は，回路基板の表面に直接はんだ付けするように設計された非常に小さい長方形の金属酸化膜抵抗器で

ある.

　可溶性抵抗器は，定格電力を超えると容易に焼損するようにつくられた巻線抵抗器である．この抵抗器は，ヒューズと抵抗の両方の機能を果たす．

In carbon and film resistors, resistors are too small to have numbers printed on them and so they are marked with some colored bands. Each color stands for a number.

The example shown in Fig1・6 represents five-band color-coded metal film resistor. This resistor has three bands for the resistance value, one multiplier and one tolerance band.

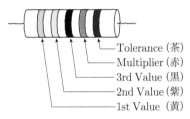

Tolerance（茶）
Multiplier（赤）
3rd Value（黒）
2nd Value（紫）
1st Value（黄）

図1・6　5 color band resistor.

● 炭素抵抗器や皮膜抵抗器では，抵抗器は小さすぎて数字を印刷できないため，いくつかの色の帯が付いている．各色は数字を表す．

　図1・6に示した例は，5本帯のカラーコード表示の金属皮膜抵抗器を表す．この抵抗器では，3本の帯で抵抗値，1本の帯で乗数，そして1本の帯で抵抗値許容差を示している．

Example　Yellow $= 4$, Violet $= 7$, Black $= 0$, Red $= 2$

$470 \times 10^2 = 47$ kΩ, Resistor tolerance(Brown) $= \pm 1$ %

例題　黄 $= 4$，紫 $= 7$，黒 $= 0$，赤 $= 2$

$470 \times 10^2 = 47$ kΩ，抵抗値許容差（茶）$= \pm 1$ %

1.6 コンデンサ

コンデンサは，電気を蓄えたり，放出したりすることができる電子部品である．コンデンサは，電気を電気エネルギーのまま蓄える．

A capacitor (condenser) is a passive element that stores electric charge. The capacitor consists of a dielectric between two conductive plates.

Capacitance is the ability of a dielectric to store an electric charge. The SI unit of capacitance is Farad (F), named after Michael Faraday.

● コンデンサは，電荷を蓄える受動素子である．コンデンサは，二つの導電板で挟まれた誘電体で構成される．

キャパシタンス（静電容量）は，誘電体が電荷を蓄える能力である．国際単位系のキャパシタンスの単位はファラド（F）で，マイケル・ファラデーにちなんで命名された．

When you apply a voltage over the two conductive plates, an electric field is created. Positive charge will collect on one plate and negative charge on the other.

The electric charge Q in Coulombs in the capacitor is defined by the following equation.

$$Q = CV \, [\mathrm{C}]$$

where V is voltage in Volts and C is capacitance in Farads.

Now differentiate both sides of the above expression with respect to time.

$$\frac{\mathrm{d}}{\mathrm{d}t} Q = \frac{\mathrm{d}}{\mathrm{d}t} CV$$

or, $i = C \dfrac{\mathrm{d}V}{\mathrm{d}t}$

From this equation, it is clear that the current flows through the capacitor only when there is a change in the voltage through the capacitor. If $\mathrm{d}V/\mathrm{d}t = 0$, the capacitor does not conduct current.

● 二つの導電板に電圧を印加すると，電界が形成される．正の電荷は一方の極板に，負の電荷は他方の極板に集まる．

クーロンを単位とするコンデンサの電荷 Q は，以下の式で定義される．

$$Q = CV \; [\mathrm{C}]$$

ここで，V は単位がボルトの電圧，C は単位がファラドのキャパシタンスである．

次に上記の式の両辺を時間で微分しなさい．

$$\frac{\mathrm{d}}{\mathrm{d}t}Q = \frac{\mathrm{d}}{\mathrm{d}t}CV$$

すなわち，$i = C \dfrac{\mathrm{d}V}{\mathrm{d}t}$

この式から，コンデンサにかかる電圧が変化したときのみ，コンデンサに電流が流れることは明らかである．$\mathrm{d}V/\mathrm{d}t = 0$ の場合，コンデンサは電流を通さない．

In the circuit shown in Fig.1・7, when the switch is closed, the capacitor is gradually charged up to the source voltage. As direct current only flows in one direction, once the capacitor is fully charged there is no more current flow.

● 図1・7に示した回路において，スイッチが閉じられると，コンデンサは

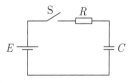

図1・7 *RC* charging circuit.

電源電圧まで徐々に充電される．直流は一方向のみに流れるので，ひとたびコンデンサが完全に充電されると，それ以上電流は流れない．

Capacitors can be fabricated onto integrated circuit chips. They are commonly used in conjunction with transistors in dynamic random access memory. The capacitors help maintain the contents of memory.

● コンデンサは，集積回路チップ上に製造することができる．これらは，ダイナミックランダムアクセスメモリのトランジスタとともに一般に用いられる．コンデンサを用いると，メモリの内容を保持することができる．

Example1 How much charge is stored in a 2 μF capacitor connected across a 50 V supply?

例題1 50 Vの電源に接続された2 μFのコンデンサに蓄えられる電荷はいくらか．

Solution(答) $Q = CV = 2 \times 10^{-6} \times 50 = 100 \times 10^{-6}$ C
$$= 100 \ \mu\text{C}$$

Example2 A constant current 5 mA charges a 10 μF capacitor for 1 s. How much is the voltage across the capacitor?

例題2 定電流5 mAで10 μFのコンデンサを1秒間充電する．コンデンサ両端の電圧はいくらか．

Solution(答) Let us find the stored charge first.

はじめに，蓄えられた電荷を求めることにする．

$$Q = It = 5 \times 10^{-3} \times 1 = 5 \times 10^{-3} \text{ C}$$

$$\therefore \ V = \frac{Q}{C} = \frac{5 \times 10^{-3}}{10 \times 10^{-6}} = 0.5 \times 10^3 = 500 \text{ V}$$

1.7 コイル

コイルは，電気が供給されると，エネルギーを磁気エネルギーの形で蓄える受動部品である．

A coil, also called an inductor, is a circuit element that stores energy in the form of magnetic field. Some coils are called air-core coils, which are wound around either air or others materials such as glass, plastic or ceramic. Others are coils that are wound around an iron core.

● コイルは，インダクタとも呼ばれ，磁界の形でエネルギーを蓄える回路素子である．コイルの中には空心コイルと呼ばれるものがある．これは，空気またはガラス，プラスチック，あるいはセラミックのような空気以外の材料の周りに巻かれているものである．ほかにも鉄心の周りに巻かれたコイルがある．

If the current through a coil is changed then the magnetic flux through the coil also changes, and this will induce voltage in itself. The property of the coil is called self inductance.

The basic unit of measurement for inductance is called the Henry (H). The inductance is directly proportional to the number of turns in the coil.

Fig.1·8 shows a graphic symbol of an air-core coil that does not use a magnetic core made of a ferromagnetic material.

● コイルを通る電流が変化すると，コイルを鎖交する磁束も変化し，これがそれ自体電圧を誘導する．コイルのこの特性は自己インダクタンスと呼ばれる．インダクタンスの基本的な測定単位は，ヘンリー（H）と呼ばれる．インダクタンスは，コイルの巻き数に正比例する．

図1·8 Air-core coil.

図1・8は，強磁性体からなる磁心を用いない空心コイルの図記号を示している．

Fig.1·9 illustrates the circuit and the current waveform of self induction.

When the switch is closed at $t = t_1$, current flows through the coil and some magnetic flux is linked with the coil. A change of magnetic flux linkage occurs when the switch is closed. This induces an electromotive force and an induced current which opposes the change in the magnetic flux.

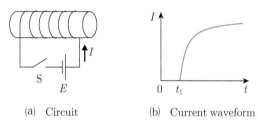

| (a) Circuit | (b) Current waveform |

図1・9 Self induction.

● 図1・9は，自己誘導の回路と電流波形を表している．
　スイッチが時刻 t_1 で閉じられると，コイルに電流が流れ，磁束がコイルを鎖交する．鎖交磁束の変化は，スイッチが閉じられたときに生じる．これにより磁束の変化を妨げる起電力と誘導電流が誘導される．

Self-induction is the production of an induced electromotive force in a conductor itself when the current through the conductor changes. The induced electromotive force is known as the back electromotive force (E) and it is directly proportional to the rate of change of the current.

$$E = -L\frac{\mathrm{d}I}{\mathrm{d}t}\ [\mathrm{V}]$$

where L is self inductance of the coil. Note that the negative sign indicates that the induced electromotive force opposes the change in current through the coil per unit time.

Therefore $L = \dfrac{E}{\mathrm{d}I/\mathrm{d}t}$

where L is measured in henry (H). One henry is the inductance needed to induce one volt by a change of current of one ampere per second.

● 自己誘導とは，導線を流れる電流が変化すると，導線自体に誘導起電力が発生することである．誘導起電力は，逆起電力（E）として知られており，電流の変化率に正比例する．

$$E = -L\frac{\mathrm{d}I}{\mathrm{d}t}\ [\mathrm{V}]$$

ここで，L はコイルの自己インダクタンスである．負の符号は，誘導起電力が単位時間当たりにコイルを流れる電流の変化を妨げる向きを示していることに注意しなさい．

したがって，$L = \dfrac{E}{\mathrm{d}I/\mathrm{d}t}$

ここで，L はヘンリー（H）で表される．1ヘンリーは，毎秒1アンペアの電流の変化によって1ボルトを誘導するのに必要なインダクタンスである．

Example How much is the self inductance of a coil that induces 50 V when its current changes at the rate of 5 A/s ?

例題 電流が毎秒5 Aで変化するとき，50 Vを誘導するコイルの自己インダクタンスはいくらか．

Solution (答) $L = \dfrac{E}{\mathrm{d}I/\mathrm{d}t} = \dfrac{50}{5} = 10 \text{ H}$

1.8 ジュールの法則

　抵抗を流れる電流によって毎秒発生する熱量は，電流の2乗と抵抗の積に比例する．これをジュールの法則という．

When an electric current flows through a resistor, the electrical energy is converted into heat energy. This is called heat effect of electric current. Moreover, the heat resulting from electric current through the resistor is called Joule heat.

The heat produced in a conductor due to flow of current in it is proportional to square of current, resistance of conductor and the time for which current flows. This is called Joule's law.

In 1841, Joule stated that when a current I is made to flow through a resistance R for time t, heat Q is produced such that

$$Q = I^2 R t$$

Moreover, Q is written as follows using voltage V between two ends of the resistor.

$$Q = VIt$$

$$Q = \frac{V^2}{R} t$$

When current, resistor, voltage and time are expressed in amperes, ohms, volts and seconds respectively, the unit of Q is the joule.

● 　抵抗を電流が流れると，電気エネルギーは熱エネルギーに変換される．これを電流の熱作用という．さらに，抵抗を電流が流れることで発生する熱は，ジュール熱と呼ばれる．

　導体を流れる電流により発生する熱は，電流の2乗，導体の抵抗，それに電流が流れる時間に比例する．これをジュールの法則という．

　1841年に，ジュールは抵抗 R を電流 I が t 時間流れると，発熱量 Q は

$$Q = I^2 R t$$

となることを示している．

　さらに，Q は抵抗の両端の電圧 V を用いて以下のように表される．

$$Q = V I t$$

$$Q = \frac{V^2}{R} t$$

電流，抵抗，電圧および時間がそれぞれアンペア，オーム，ボルトおよび秒で表された場合，Q の単位はジュールである．

Example1　An electric heater having resistance equal to 5 Ω is connected to electric source. If it produces 180 J of heat in one second, find the potential difference between the two terminals of the electric circuit.

例題1　5 Ωの抵抗を有する電気ヒータが電源に接続されている．このヒータが毎秒180 Jの熱を発生するとすれば，この電気回路の端子間の電位差を求めなさい．

Solution（答）　Joule's law is given by

$$Q = I^2 R t$$

ジュールの法則は，$Q = I^2 R t$ で与えられる．

$$180 \text{ J} = I^2 \times 5 \text{ Ω} \times 1 \text{ s}$$

$$I^2 = 180/5 = 36$$

$$\therefore \ I = 6 \text{ A}$$

$$V = IR = 6 \text{ A} \times 5 \text{ Ω} = 30 \text{ V}$$

Example2　A direct-current generator has an electromotive force of 100 V and provides a current of 8 amps. How much energy does it provide each minute?

例題2 起電力が 100 V で，8 A の電流を供給する直流発電機がある．毎分どれだけのエネルギーを供給できるか．

Solution (答) Joule's law is given by

$$Q = I^2Rt = VIt$$

ジュールの法則は，$Q = I^2Rt = VIt$ で与えられる．

$$Q = 100 \text{ V} \times 8 \text{ A} \times 60 \text{ s} = 48 \text{ kJ}$$

1.9 ボルタ電池

電池の歴史はボルタ電池（Voltaic cell）に始まる．イタリアの物理学者 Alessandro Volta（1745-1827）が 1800 年に発明したもので，現在も多くの電池がボルタ電池の原理を採用している．電圧の単位として用いられるボルト [V] は，ボルタにちなむ．

As shown in Fig.1・10, one zinc plate and one copper plate are immersed inside an electrolyte of dilute sulfuric acid.

First, let us consider the case where the switch S is open. Since zinc has higher ionization tendency than copper, zinc atoms are converted into zinc ions (Zn^{2+}) and come out in the electrolyte. Therefore, the zinc plate is negatively charged.

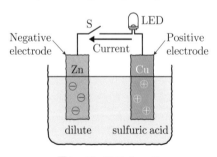

図 1・10　Voltaic cell.

The dilute sulfuric acid is ionized into two positively charged hydrogen ions ($2H^+$) and one negatively charged sulfate ion (SO_4^{2-}).

The ionized positive hydrogen ions ($2H^+$) reach the copper electrode and combine with two electrons to form hydrogen gas that is evolved at the copper electrode ($2H^+ + 2e^- \rightarrow H_2$). Therefore, the copper plate is positively charged.

Next, let us consider the case where the switch S is closed. Once the copper electrodes starts drawing electrons through the conducting wire, a chemical reaction helps to keep the current going. Every zinc atom that loses electrons to the copper electrode becomes a zinc ion with a double positive charge (Zn^{2+}). Sulfate ions promptly attract the zinc ions into the solution, where they combine to form zinc sulfate ($Zn^{2+} + SO_4^{2-} \rightarrow ZnSO_4$).

The chemical reaction taking place in a voltaic cell can be summarized as follows.

At zinc plate: $\qquad Zn \rightarrow Zn^{2+} + 2e^-$

In the solution: $\quad H_2SO_4 \rightarrow 2H^+ + SO_4^{2-}$

$\qquad\qquad\qquad\quad Zn^{2+} + SO_4^{2-} \rightarrow ZnSO_4$

At copper plate: $\ 2H^+ + 2e^- \rightarrow H_2$

Voltaic cell uses chemical reactions to produce electrical energy. It is also called a galvanic cell, named after Luigi Galvani (1737-1798).

● 図1・10に示したように，1枚の亜鉛板と1枚の銅板が希硫酸からなる電解液に浸されている．

はじめに，スイッチSが開いている場合を考える．亜鉛は銅よりイオン化傾向が大きいので，亜鉛の原子は亜鉛イオン（Zn^{2+}）となり，電解液中に溶けだす．それゆえに，亜鉛板は負に帯電する．

希硫酸は，2個の正に帯電した水素イオン（2H⁺）と1個の負に帯電した硫酸イオン（SO_4^{2-}）に電離される．

イオン化された正の水素イオンは銅電極に到達し，2個の電子と結合して銅電極で発生する水素ガスを生成する（$2H^+ + 2e^- \rightarrow H_2$）．したがって，銅板は正に帯電する．

次に，スイッチSが閉じている場合を考える．いったん銅電極が導線を通して電子を引きつけると，化学反応により電流が流れ続ける．銅電極に到達する電子を失ったどの亜鉛原子も，2価の正電荷をもつ亜鉛イオンになる．溶液中では，硫酸イオンがすぐに亜鉛イオンを取り込み，そこでそれらは結合し，硫酸亜鉛（$Zn^{2+} + SO_4^{2-} \rightarrow ZnSO_4$）を生成する．

ボルタ電池で起こる化学反応は，以下のように要約することができる：

亜鉛板：$Zn \rightarrow Zn^{2+} + 2e-$

電解液：$H_2SO_4 \rightarrow 2H^+ + SO_4^{2-}$

$Zn^{2+} + SO_4^{2-} \rightarrow ZnSO_4$

銅　板：$2H^+ + 2e^- \rightarrow H_2$

ボルタ電池は，化学反応から電気エネルギーをつくりだすために用いる．ボルタ電池はまた，ルイージ・ガルヴァーニ（1737-1798）にちなんでガルバニ電池とも呼ばれる．

1.10　電池の種類

電池には，乾電池などの放電すると再生できない一次電池と，充電すると再生できる二次電池，そして化学反応により発生した電気を継続的に取り出す燃料電池がある．

A battery generally consists of a positive electrode, a negative electrode, and an electrolyte. The chemical reaction between the electrodes and the electrolyte results in a separation of ions and electrons. This causes a potential difference between the electrodes.

● 電池は，一般に，正極，負極および電解液で構成される．その電極と電解液間で起こる化学反応は，イオンと電子に分離をもたらす．これが起因

となって，電極間に電位差が生じる．

Today, we rely on batteries as a power source for many electronic products. For example, the batteries are used in cars, smartphones, hand-held radios, notebook computers, cameras, and tablet terminals.

When choosing a battery, a lot of things must be considered. A wide variety of shapes, sizes, and ratings are necessary to meet the requirement of an enormous number of applications.

● 　今日，私たちは多くの電子製品の電源として電池に頼っている．例えば，自動車，スマートホン，携帯ラジオ，ノートパソコン，カメラ，そしてタブレット端末に電池が用いられている．
　　電池を選ぶときには，多くのことを考慮しなければならない．膨大な数のアプリケーションの要求を満たすには多種多様な形状，大きさおよび定格が必要である．

Some batteries cannot be recharged once they are worn out. Others can be recharged hundreds or even thousands of times before they are no longer able to produce or maintain the rated output voltage.

There are two types of batteries, primary and secondary. A primary battery cannot be recharged, whereas a secondary battery can be recharged because the chemical reaction is reversible.

● 　電池のなかには，使い切ったら再充電できないものがある．ほかのものは，定格出力電圧を生成または維持することができなくなるまで，数百回または数千回も再充電することができる．
　　一次と二次の2種類の電池がある．一次電池は再充電できないのに，二次電池は化学反応が可逆的であるという理由で再充電できる．

（ⅰ）　アルカリ電池（Alkaline Battery）

Alkaline battery has much higher capacities and longer shelf life compared to zinc battery. The main operation principle of the alkaline battery is based on the reaction between zinc (Zn) and manganese dioxide (MnO_2). Alkaline battery is so named because the electrolyte used in it is potassium hydroxide, a purely alkaline substance.

● アルカリ電池は，亜鉛電池と比べてはるかに高い容量と長寿命を有する．アルカリ電池の主な動作原理は，亜鉛（Zn）と二酸化マンガン（MnO_2）との反応に基づく．アルカリ電池は，それに使用される電解液が純粋にアルカリ性の物質である水酸化カリウムであるため，その名前が付けられている．

(ii) 酸化銀電池（Silver Oxide Cell）

The silver oxide cell has good shelf life, and ability to operate over a wide temperature range. Furthermore, since it has a characteristic of maintaining voltage up to the end of its life, it is used widely in electronic equipment that requires a small compact power source. The most common silver oxide cell is the small button cell used in cameras, watches, and calculators.

● 酸化銀電池は，保存期間が長く，広い温度範囲で動作が可能である．さらに，酸化銀電池は，寿命に至るまで電圧を維持する特性があるので，小形の電源を必要とする電子機器に広く用いられている．最も一般的な酸化銀電池は，カメラ，腕時計，電卓に使用される小形のボタン電池である．

(iii) 鉛蓄電池（Lead-Acid Battery）

Lead-acid battery is a high-capacity rechargeable battery and widely used in our lives. For example, it is used to supply energy to lighting systems and accessories within a vehicle. Also, it is widespread use as a backup power supply in a hospital during power failure.

● 鉛蓄電池は，大容量の充電式電池であり，私たちの生活のなかで広く使

用されている．例えば，鉛蓄電池は，自動車内の照明システムおよびアクセサリーにエネルギーを供給するために使用される．また，停電時に病院内のバックアップ電源として広く利用されている．

(iv)　リチウムイオン電池（Lithium-Ion Battery）

Lithium-ion battery is extremely popular and has found widespread use in today's consumer products such as a smartphone, and a notebook computer. This battery is now being used in place of a lead-acid battery and a NiMH battery.

● 　リチウムイオン電池は，非常に普及しており，スマートホンやノートパソコンのような今日の家電製品において広く使用されている．この電池は，今や鉛蓄電池やニッケル水素電池に代わって用いられている．

(v)　太陽電池（Solar Battery）

Solar battery is an electronic device that catches sunlight and converts it directly into electricity. In most applications, the solar battery is used in combination with a lead-acid battery. In the presence of sunlight, the solar battery charges the secondary battery and supplies power to the load. Power cables are not necessary in traffic signs using solar battery.

● 　太陽電池は，太陽光を捉えて電気に直接変換する電子装置である．ほとんどのアプリケーションで，太陽電池は鉛蓄電池とともに用いられる．太陽光がある場合には，太陽電池は二次電池を充電し，負荷に電力を供給する．電源ケーブルは，太陽電池を用いた交通標識では必要としない．

(vi)　燃料電池（Fuel Cell）

Fuel cell converts the chemical energy of a fuel directly into electrical energy by chemical reaction. It is an electrochemical energy conversion device that produces electricity, water, and heat.

A fuel cell looks like a battery in many respects, but it can supply electrical energy over a much longer period of time. This is because a fuel cell is continuously supplied with fuel and oxygen from an external source.

Unlike conventional vehicle which runs on gasoline or diesel oil, a fuel cell vehicle drives the electric motor using electricity produced by reacting hydrogen and oxygen.

● 燃料電池は，化学反応によって燃料の化学エネルギーを直接電気エネルギーに変換する．燃料電池は，電気，水および熱を生成する電気化学的エネルギー変換装置である．

　燃料電池は，多くの点で電池に似ているが，はるかに長時間にわたって電気エネルギーを供給することができる．これは，燃料電池には外部から燃料と酸素が連続的に供給されるからである．

　ガソリンやディーゼルで走行する従来の車両とは異なり，燃料電池自動車は水素と酸素を反応させて発電した電気を用いて電気モータを駆動する．

電気回路の基礎

電気回路は電源と抵抗，コンデンサ，コイルなどの受動素子（線形素子）で構成される．この章では，電気回路の抵抗，電流，電圧の関係など電気を理解するための基本法則について述べる．

2.1 オームの法則

電気回路に電流が流れるときの電流の大きさと電圧の関係を調べたもので，電流の大きさは電圧に比例し，抵抗に反比例するというのがオームの法則である．

The relationship between voltage, current and resistance in an electric circuit was discovered in 1826 by the German physicist Georg Simon Ohm. He found that the electric current flowing through a resistance is directly proportional to the voltage and inversely proportional to the resistance. The relationship between the voltage, current and resistance is called Ohm's law.

In Fig.2・1, Ohm's law states that the current I, is directly proportional to the voltage V, and inversely proportional to the resistance R.

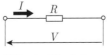

図2・1　Ohm's law.

Ohm's law can be expressed as follows.

$$I = \frac{V}{R}$$

where I = current in amps, V = voltage in volts, and R = resistance in ohms.

The equation for calculating the voltage or resistance can be expressed as follows.

$$V = IR$$

$$R = \frac{V}{I}$$

● 電気回路における電圧，電流，そして抵抗の関係は，ドイツの物理学者ゲオルク・ジーモン・オームによって，1826年に発見された．彼は抵抗を流れる電流が，電圧に比例し，抵抗に反比例することを発見した．この電圧，電流，そして抵抗の関係はオームの法則と呼ばれる．

図2・1において，オームの法則は，電流 I は電圧 V に比例し，抵抗 R に反比例する．オームの法則は，以下のように表される．

$$I = \frac{V}{R}$$

ここで，I はアンペアを単位とする電流，V はボルトを単位とする電圧，R はオームを単位とする抵抗である．

電圧または抵抗を求める計算式は，次のように表すことができる．

$$V = IR$$

$$R = \frac{V}{I}$$

In Fig.2・2, an LED emits light when an electric current passes through it. The resistance R is a ballast resistor used to limit the current through the LED. The resistance of the ballast resistor is calculated as follows.

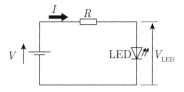

図2・2 LED lighting circuit.

$$R = \frac{V - V_{\text{LED}}}{I}$$

where V is the electromotive force of a source. V_{LED} is the LED voltage, and I is the electric current.

In this Fig., let us calculate R when $V_{\text{LED}} = 2$ V, $I = 10$ mA, and $V = 6$ V.

$$R = \frac{6 - 2}{0.01} = 400 \ \Omega$$

● 図2・2で，LEDは電流が流れると発光する．抵抗 R はLEDを流れる電流を制限するために用いられる安定抵抗器である．安定抵抗器の抵抗は，以下のように計算される．

$$R = \frac{V - V_{\text{LED}}}{I}$$

ここで，V は電源の起電力である．V_{LED} はLEDの電圧，I は電流である．

この図において，$V_{\text{LED}} = 2$ V，$I = 10$ mA，それに $V = 6$ V として R を計算すると

$$R = \frac{6 - 2}{0.01} = 400 \ \Omega$$

2.2 合成抵抗

複数の抵抗を接続した場合，全体としての抵抗を合成抵抗（combined resistance）という．抵抗の接続の代表的な方法として，直列接続，

並列接続，そしてそれらを組み合わせた直並列接続がある.

(i) 直列接続

　図2・3に示したように，抵抗 R_1，R_2 の二つの抵抗に同じ電流が流れる接続を直列接続という.

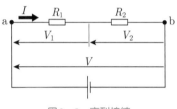

図2・3　直列接続

　この回路に流れる電流 I [A] は，そのまま抵抗 R_1 [Ω] と R_2 [Ω] を流れる．それぞれの抵抗に生じる電圧を V_1 [V]，V_2 [V] とすると，オームの法則から次の関係が成り立つ.

$$V_1 = IR_1, \quad V_2 = IR_2$$

端子a-b間の電圧 V [V] は，V_1 と V_2 の和に等しく，次のように表される.

$$V = V_1 + V_2 = IR_1 + IR_2$$

したがって，合成抵抗を R [Ω] とすると，

$$R = \frac{V}{I} = R_1 + R_2$$

となる．すなわち，直列接続の合成抵抗は，各抵抗の和に等しい.

　When two resistors are daisy chained in a single line, they are said to be connected in series. A current flowing through the first resistance also passes through the second resistance. Therefore, resistances in series have a common current.

　Any series circuit is a voltage divider in which the voltage drop across individual resistance is proportional to the series resistance values.

● 二つの抵抗が1本の線に数珠つなぎになっているとき，それらは直列接続という．最初の抵抗を流れる電流は，2番目の抵抗をも流れる．したがって，直列接続の抵抗には共通の電流が流れる．

抵抗を直列に接続したどんな直列回路も，各抵抗での電圧降下はその抵抗値に比例するという点で，電圧分圧器となる．

(ii) 並列接続

Fig. 2·4 shows a circuit consisting of two resistances connected in parallel. It is clear that the voltage drop V across the two resistances is the same. According to Ohm's law, the currents flowing through the individual resistances are $I_1 = \dfrac{V}{R_1}$ and $I_2 = \dfrac{V}{R_2}$.

Conservation of charge implies that the total current I is equal to the sum of these currents I_1 and I_2.

● 図2·4は並列に接続された二つの抵抗からなる回路を示す．二つの抵抗の両端の電圧降下 V は，同じであることがわかる．オームの法則から，それぞれの抵抗を流れる電流は，$I_1 = \dfrac{V}{R_1}$，$I_2 = \dfrac{V}{R_2}$ となる．

電荷保存則から，全電流 I は，電流 I_1 と I_2 の和に等しい．

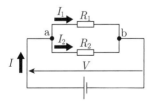

図2·4 parallel connection.

また，二つの抵抗の両端の電圧は等しく，V [V] なので，I [A] は次のようになる．

$$I = I_1 + I_2 = \frac{V}{R_1} + \frac{V}{R_2} = \left(\frac{1}{R_1} + \frac{1}{R_2} \right) V$$

したがって，合成抵抗を R [Ω] とすると，

$$\frac{1}{R} = \frac{I}{V} = \frac{1}{R_1} + \frac{1}{R_2}$$

となり，R [Ω] は以下のように表される．

$$R = \frac{1}{\dfrac{1}{R_1} + \dfrac{1}{R_2}} = \frac{R_1 R_2}{R_1 + R_2}$$

抵抗の並列接続では，各抵抗の両端の電圧は等しい．一方，回路を流れる電流は抵抗の大きさによって分かれて流れ，これを分流と呼んでいる．

A series-parallel circuit is a circuit that combines both series and parallel connections. When analyzing the series-parallel circuit, the individual laws of series and parallel circuits can be applied to produce a much simpler overall circuit.

● 直列列回路は，直列接続と並列接続の両方を組み合わせた回路である．直並列回路を解析する際，直列回路や並列回路に関する個々の法則を適用することで，より単純な全体の回路をつくり出すことができる．

2.3 電源の内部抵抗

実際の電源は，抵抗の大きさによって電圧が変化したり，電流が変化したりする．このような変化は，電源に内部抵というものがあり，その影響によって説明される．電圧源では，電圧源と内部抵抗

は直列の関係にある．一方，電流源では，電流源と内部抵抗は並列の関係がある．

図2・5に示すように，電池は起電力 E [V] と電池自身がもっている内部抵抗 r [Ω] の直列回路と考えられる．この回路で，スイッチS が開いているときは回路に電流が流れないので，電圧計は電池の端子電圧 V [V]，すなわち電池の起電力 E [V] を示している．

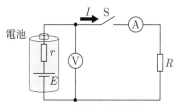

図2・5　電池の内部抵抗

次に，スイッチSを閉じると回路に電流が流れ，このとき電流計の指示を I [A] とする．一方，電圧計の指示を V [V] とすると，$E - V$ は電流 I が流れたことによる電池の内部抵抗 r の両端の電圧を表す．

The voltage that appears at the battery terminals is less than the voltage produced by the voltage source by an amount equal to the voltage dropped across the internal resistance. Therefore, the output voltage produced by the battery when connected to a load of R is given as follows.

$$V = E - Ir \tag{2-1}$$

where I is the current flowing through the load, and r is the internal resistance.

● 電池の両端の電圧は，内部抵抗による電圧降下に等しい分だけ電圧源によって生み出される電圧より小さい．それゆえ，負荷 R を接続したとき電

池によって生じる出力電圧は次のように与えられる.

$$V = E - Ir \qquad (2\text{-}1)$$

ここで,Iは負荷を流れる電流,rは内部抵抗である.

電池は自分自身がもっている内部抵抗によって,起電力の一部を消費する.一般に,電池の内部抵抗rは電池の劣化が進むほど大きな値となり,その分負荷Rにかかる電圧は低下する.電池は使用すればするほど電圧が下がることはよく知られているが,電池が減ってきたというのは,化学的な特性が変化することによって次第にエネルギーが取り出せなくなることをいう.つまり,内部抵抗が次第に増加することで,取り出せる電流が減少する.

For a chemical cell, internal resistance r is mainly the resistance of the electrolyte. For a good cell, r is very low. As the cell deteriorates, r increases.

The output voltage of a cell depends on the elements used for the electrodes and the electrolyte. The current rating depends mostly on the physical size. Larger battery can supply more current.

● 化学電池において,内部抵抗rは主に電解液の抵抗である.良好な電池では,rは十分小さい.電池は劣化するにつれ,rは増加する.
電池の出力電圧は,電極や電解液に用いられている構成要素に依存する.定格電流は,主に物理的な大きさによって決まる.電池が大きくなればなるほど,より多くの電流が供給できる.

(2-1)式において,回路に流れる電流I [A] は次式で与えられる.

$$I = \frac{E}{r + R}$$

したがって,(2-1)式は次式のように表すことができる.

$$V = E - Ir = E - \frac{r}{r + R} E$$

この式から，端子電圧 V は内部抵抗 r の大きさによって変動することがわかる．特に，r が R と比較して十分小さい（$r \ll R$）場合には，r の影響を無視でき，E と V はほぼ等しくなる．

2.4 分流器と直列抵抗器

分流器と直列抵抗器は，それぞれ電流と電圧の測定範囲を広げるために回路に接続する抵抗のことである．分流器は，電流計に並列，直列抵抗器は，電圧計に直列に接続する．

（i）分流器

電流計には測定できる最大値が定まっている．しかし，1 A までしか測れない電流計で 10 A まで測定するにはどのようにすればよいか．この場合，電流計に 1 A を超える電流を流せば電流計は破損してしまうので，電流計と並列に接続した抵抗に $10 - 1 = 9$ A の電流を流し，全体として 10 A の電流を流すようにすればよい．図 2・6 に示すように，電流計の最大測定範囲を拡大するために並列に接続した抵抗を分流器という．

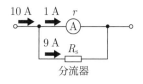

図 2・6　分流器

この図のように，電流計 A（内部抵抗 r）と並列に抵抗 R_s を接続した回路では，次の式が成り立つ．

$$1 \times r = 9 \times R_\mathrm{s}$$

よって，$R_\mathrm{s} = \dfrac{1}{9} r$ が得られ，電流計の内部抵抗の 1/9 の抵抗をも

つ分流器を用いれば，1 A の電流計で10 A まで測れることになる．
すなわち，電流計の読みを10倍したものが全体の電流となる．

　一般に，測定可能な電流の n 倍の電流を測定するには，

$R_s = \dfrac{r}{n-1}$ の抵抗をもつ分流器を用いればよい．ここで，n を分流
器の倍率と呼ぶ．

A shunt is a useful means of extending the current range of an ammeter. As current divides between two resistances in parallel it is possible to increase the range of a microammeter or milliammeter by paralleling an additional resistance with the inherent resistance of the meter itself. This is called a shunt.

● 　分流器は，電流計の測定範囲を広げるのに役に立つ方法である．抵抗の
並列接続では電流の分流が起こるので，電流計自身の内部抵抗と並列に抵
抗を付加することで，マイクロアンペア計やミリアンペア計の測定範囲を
拡大させることができる．これは分流器と呼ばれる．

Example In Fig.2·7, calculate the value of the shunt resistance R_s required to extend the range of the meter movement to 100 μA.

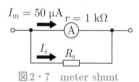

図2·7　meter shunt

例題 図2·7において，電流計の測定範囲を 100 μA に拡大するために必要
な分流器の抵抗値を計算せよ．

Solution (答) 　$I_m r = I_s R_s$

$$\therefore \quad R_s = \frac{I_m r}{I_s} = \frac{50 \times 10^{-6} \times 1 \times 10^3}{50 \times 10^{-6}} = 1 \times 10^3 \ \Omega$$

$$= 1 \text{ k}\Omega$$

(ii) 直列抵抗器

電流計の測定範囲を拡大する分流器と同様に，電圧計でも直列抵抗器（以前は倍率器と呼ばれていた）を用いて電圧の測定範囲を拡大することができる．

電圧計に直列に直列抵抗器を接続すると，測定電圧が直列抵抗器の抵抗と電圧計の内部抵抗により分圧されるので，結果的に電圧計の最大目盛よりも大きな電圧を測定できる．測定した数値は，直列抵抗器の倍率によって読み替えを行う．

What is the purpose of a series resistor used with a voltmeter?

It is used to increase the voltage-indicating range of the voltmeter.

● 電圧計とともに用いられる直列抵抗器の目的は何か？
　直列抵抗器は，電圧計の測定範囲を拡大させるために用いられる．

図2・8に直列抵抗器の構成を示す．ここで，V_r [V] は電圧計本体の最大目盛，r [Ω] は電圧計の内部抵抗，V [V] は測定範囲を拡大したときの最大目盛，R_m [Ω] は直列抵抗器の抵抗である．

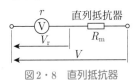

図2・8　直列抵抗器

図において，V_r [V] は V [V] を r [Ω] と R_m [Ω] の比で分割したもので，次のように表される．

$$V_r = \frac{r}{R_m + r} V$$

また，測定範囲を拡大したときの最大目盛 V [V] と，電圧計に加わる電圧 V_r [V] との比を m とすると，m は次のように表される．

$$m = \frac{V}{V_{\mathrm{r}}} = \frac{R_{\mathrm{m}} + r}{r} = \frac{R_{\mathrm{m}}}{r} + 1$$

この式から $R_{\mathrm{m}}\,[\Omega]$ は，

$$R_{\mathrm{m}} = r(m - 1)$$

となる．したがって，この R_{m} を用いれば測定範囲を m 倍にすることができ，m を直列抵抗器の倍率という．例えば，最大目盛 50 V，内部抵抗 1 kΩ の電圧計に，直列抵抗器 $R_{\mathrm{m}}\,[\mathrm{k}\Omega]$ を接続して最大目盛 500 V の電圧計をつくる場合，R_{m} の値をいくらにすればよいか求めてみよう．まず，倍率 m を計算すると，$500/50 = 10$ となる．次に，R_{m} の式に $m = 10$，$r = 1$ kΩ を代入して計算すると，$R_{\mathrm{m}} = 9$ kΩ が得られる．

In order to prevent the meter movement of a voltmeter from drawing too much current from the circuit, it is necessary to connect a resistor, called a series resistor, in series with that voltmeter. The series resistor acts to limit the current through the movement, ensuring that it does not exceed the coil's full-scale deflection current.

● 電圧計の可動部に回路から非常に多くの電流が流れ込まないようにするために，電圧計と直列に直列抵抗器と呼ばれる抵抗を接続することが必要である．直列抵抗器は，電圧計の可動部に流れる電流を制限するように働き，その結果，目盛の最大振れ幅を超える電流が，コイルに流れないことを確実にしている．

2.5 電気抵抗

電気抵抗または単に抵抗とは，電流の流れにくさを表すものである．電子が導体内部を移動する際，導体の原子との間で衝突を引き起こすが，このような衝突が多くなると電子の流れが妨げられる．

この電子の流れを妨げる度合いを抵抗と呼んでいる．電気を通しやすい導体内では，自由電子が多いほど電流が流れやすく，少ないほど電流が流れにくい．

抵抗は，導体の種類や，断面積，長さによって変化するだけでなく，周囲の温度によってもその値が変わる．その大きさを統一的に表すのに，抵抗率（resistivity）を用いる．抵抗率は断面積が1 m^2，長さが1 mの導体の抵抗値で定義される．したがって，抵抗率ρがわかれば，断面積S $[m^2]$，長さl $[m]$の導体の抵抗R $[\Omega]$は次の式から求められる．

$$R = \rho \frac{l}{S}$$

ここで，抵抗率ρの単位はオームメートル$[\Omega\cdot m]$である．

Longer the length of the conductor, greater will be the resistance of the conductor. Moreover, greater the diameter of the conductor, lower will be the resistance.

● 導体の長さが長くなればなるほど，その抵抗値はますます大きくなる．さらに，その直径が大きくなればなるほど，抵抗値はますます小さくなる．

一般に，金属などの導体では，温度が高くなると金属導体の原子の振動が激しくなり，自由電子との衝突が増加する．その結果，自由電子の移動が妨げられ，電気抵抗は増加する．一方，シリコンなど半導体では，温度が上昇すると自由電子などの電気の運び手が増加するため，電流が流れやすくなる．このため，ほとんどの場合，温度の上昇に伴い抵抗は低下する．

電流の流れを妨げる程度を表すのが抵抗率であるが，反対に電流の流れやすさの程度を表すのに導電率（conductivity）がある．導電率の量記号にはσ，単位記号にはジーメンス毎メートル$[S/m]$を用い

る．導体に関しては，導電率で考えた方が便利な場合がある．導電率 σ と抵抗率 ρ とは逆数の関係がある．

$$\sigma = \frac{1}{\rho}$$

また，パーセント導電率もよく用いられる．これは，電流の流れやすさを表す方法の一つとして，国際標準軟銅の導電率 σ_S [S/m] を基準にして，他の材料の導電率 σ [S/m] との比を次式のように百分率で表したものである．

$$\text{パーセント導電率} = \frac{\sigma}{\sigma_S} \times 100 \, [\%]$$

$$(\sigma_S = 1.724 \ 1 \times 10^{-2} \, \mu\Omega \cdot m)$$

Resistance of any material varies with temperature. For temperature range is not too great, this variation can be represented approximately as a linear relation

$$R_T = R_t \, [1 + \alpha \, (T - t)]$$

where R_T and R_t are the resistance values at temperature T and t, respectively. t is the room temperature (usually specified at 20 ℃). α is the temperature coefficient of resistance. Pure metals have a small, positive value of α, which means that their resistance increases with increasing temperature. From temperature measurements of R you can find α. There are materials in which resistance decreases with increasing temperature. A thermistor is an example of such a material.

● どのような物質も，その抵抗は温度によって変化する．温度範囲がさほど大きくない場合，この変化は近似的に線形関係

$$R_T = R_t[1 + \alpha \, (T - t)]$$

として表すことができる．ここで，R_T と R_t はそれぞれ温度 T と t での抵抗値である．温度 t は室温（通常は20 ℃）である．α は抵抗温度係数である．この α は，純粋な金属では小さく，温度が上昇するにつれて抵抗が増加する正の値を有している．温度変化に対する抵抗を測定することで，α は求められる．物質のなかには，温度の上昇に対して抵抗が減少するものもある．サーミスタは抵抗温度係数が負の物質の例である．

2・6　キルヒホッフの法則

　キルヒホッフの法則とは，回路中の任意の電流の分岐点において，流れ込む電流の和と流れ出る電流の和が等しいという電流則（第1法則）と，回路の任意の閉回路において，起電力と負荷での電圧降下の和が等しいという電圧則（第2法則）からなる．

Kirchhoff's current and voltage laws were stated in 1847 by the German physicist Gustav Robert Kirchhoff. All circuits can be solved by Kirchhoff's laws because the laws do not depend on series or parallel connections.

● 　キルヒホッフの電流則と電圧則は，ドイツの物理学者グスタフ・ロベルト・キルヒホッフによって1847年に発表された．キルヒホッフの法則は，直列または並列接続に依存しないので，どのような回路もこれらの法則を用いて解くことができる．

(ⅰ)　キルヒホッフの電流則（Kirchhoff's Current Law：KCL）

The algebraic sum of the currents entering and leaving any node in a circuit must equal zero.

● 　回路中の任意の節点に流入する電流と流出する電流の代数和はゼロである．

　例えば，図2・9の回路の節点Oにおいて，流入する電流を I_1，I_2，流出する電流を I_3 とすると，電流の連続性から節点に流入する電流は流出する電流に等しいので次の式が成り立つ．

$$I_1 + I_2 = I_3$$

流入する電流を正，流出する電流を負で表現すると次式となる．

$$I_1 + I_2 - I_3 = 0$$

　これをキルヒホッフの電流則またはキルヒホッフの第1法則という．また，このような式を，キルヒホッフの電流則に基づいた電流方程式または節点方程式という．

The currents entering and leaving node O in Fig.2･9 can be compared to the flow of water in a pipeline. The total of all waters entering node O must equal the total of all waters leaving node O.

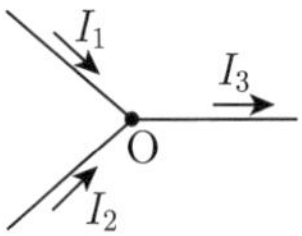

図2･9　キルヒホッフの電流則の例

● 　図2･9において，接続点Oに流入・流出する電流は，配管を流れる水に例えられる．接続点Oに流入する水の総量は，接続点Oから流出する水の総量に等しくなければならない．

(ii) キルヒホッフの電圧則 (Kirchhoff's Voltage Law：KVL)

The algebraic sum of the voltage sources and IR voltage drops in any closed path must equal zero.

● 　任意の閉回路において，起電力と抵抗に生じる電圧降下の代数和はゼロである．

　図2･10に示すように，回路に流れる電流の向きを矢印のように仮定する．破線で示した閉回路の向きと一致する電圧降下および起電力は正とし，逆向きの場合は負として扱う．

　図2･10の閉回路Iについて，電流 I_1，I_2，I_3 は各枝を流れる枝電

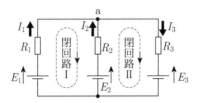

図2・10　キルヒホッフの電圧則の例

流と呼ばれる．起電力の総和は $E_1 - E_2$，電圧降下の総和は $I_1 R_1 - I_2 R_2$ である．E_2 の負の符号は，任意に決めた破線の向きと起電力の向きが逆であることを表している．また，$I_2 R_2$ の負の符号は，破線の向きと電圧降下の向きが逆であることを表している．

キルヒホッフの電圧則により，起電力の総和と電圧降下の総和が等しいので，次の式が成り立つ．

$$E_1 - E_2 = I_1 R_1 - I_2 R_2$$

同様に，閉回路IIについては次の式が成り立つ．

$$E_2 - E_3 = I_2 R_2 + I_3 R_3$$

これらの式を，キルヒホッフの電圧則に基づいた電圧方程式または閉路方程式という．

次に，節点aにおいて，キルヒホッフの電流則を適用すると，次の式が成り立つ．

$$I_3 = I_1 + I_2$$

キルヒホッフの電流則と電圧則を用いると，起電力と抵抗が与えられれば，その回路網を流れる電流と各抵抗における電圧降下を求めることができる．なお，キルヒホッフの電圧則はキルヒホッフの第2法則ともいう．

Kirchhoff's voltage and current laws can be applied to all types of electronic circuits, not just those containing dc voltage sources and resistors. For example, KVL and KCL can be applied when analyzing circuits containing diodes, transistors, operational amplifiers, etc.

● キルヒホッフの電圧則と電流則は, 直流電圧源や抵抗を含むものだけでなく, あらゆるタイプの電子回路に適用できる. 例えば, ダイオードやトランジスタ, オペアンプなどからなる回路を解析する際にも, このキルヒホッフの電圧則と電流則を当てはめることができる.

2.7 重ね合わせの理

　重ね合わせの理とは, 複数の電源からなる回路において, 任意の点における電流および任意の点の間の電圧は, 各電源が単独に存在する場合の電圧および電流の和に等しいというものである.

The superposition theorem is a way to determine the currents and voltages present in a network that has multiple sources.

The superposition theorem states that the total current in any part of a network equals the algebraic sum of the currents produced by each source acting separately.

To use one source at a time, all other sources are turned off temporarily. We replace every voltage source by a short circuit, and every current source by an open circuit. By using this method, a simpler and more manageable circuit can be obtained.

● 重ね合わせの理は, 複数の電源を有する回路網に存在する電流および電圧を決定する方法である.
　重ね合わせの理は, 回路網の任意の部分における全電流は, それぞれの電源が単独で存在した場合に生成される電流の代数和に等しいことを示している.
　　一度に一つの電源を用いるために, それ以外のすべての電源は一時的に

遮断される．各電圧源は短絡，電流源は開放する．この方法を用いることで，より単純な扱いやすい回路が得られる．

図 2・11 の回路の枝電流 I_1，I_2，I_3 を，重ね合わせの理を用いて求める．

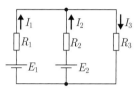

図 2・11　重ね合わせの理を適用する回路

この回路は，電源が二つあるので，一つの電源のみを含んだ二つの回路（図 2・12(a) と(b)）に分解して考える．この例では，どちらも電圧源なので，回路を分解する際，その部分を短絡する．

図 2・12(a)，(b)に示すように枝電流を設定すると，I_1，I_2，I_3 は次のように表される．

$$I_1 = I_1{}' + (-I_1{}'')$$
$$I_2 = (-I_2{}') + I_2{}''$$
$$I_3 = I_3{}' + I_3{}''$$

図 2・12(a)の電源 E_1 のみを含む回路において，キルヒホッフの電流則・電圧則を適用すると，

$$I_1{}' = I_2{}' + I_3{}' \tag{2-2}$$

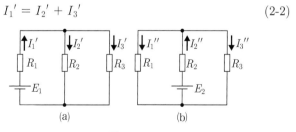

(a)　　　　　　　　　(b)

図 2・12

$$E_1 = I_1'R_1 + I_2'R_2 \tag{2-3}$$

$$E_1 = I_1'R_1 + I_3'R_3 \tag{2-4}$$

(2-2)式を(2-3)式と(2-4)式にそれぞれ代入すると,

$$E_1 = I_2'(R_1 + R_2) + I_3'R_1$$

$$E_1 = I_2'R_1 + I_3'(R_1 + R_3)$$

となる. これを I_2' および I_3' について解くと,

$$I_2' = \frac{R_3 E_1}{R_1 R_2 + R_2 R_3 + R_3 R_1}$$

$$I_3' = \frac{R_2 E_1}{R_1 R_2 + R_2 R_3 + R_3 R_1}$$

$$\therefore \quad I_1' = I_2' + I_3' = \frac{(R_2 + R_3)E_1}{R_1 R_2 + R_2 R_3 + R_3 R_1}$$

次に, 図2・12(b)の電源 E_2 のみを含む回路において, キルヒホッフの電流則・電圧則を適用すると,

$$I_2'' = I_1'' + I_3'' \tag{2-5}$$

$$E_2 = I_2''R_2 + I_1''R_1 \tag{2-6}$$

$$E_2 = I_2''R_2 + I_3''R_3 \tag{2-7}$$

(2-5)式を(2-6)式と(2-7)式にそれぞれ代入すると,

$$E_2 = I_1''(R_1 + R_2) + I_3''R_2$$

$$E_2 = I_1''R_2 + I_3''(R_2 + R_3)$$

となる. これを I_1'' および I_3'' について解くと,

$$I_1'' = \frac{R_3 E_2}{R_1 R_2 + R_2 R_3 + R_3 R_1}$$

$$I_3'' = \frac{R_1 E_2}{R_1 R_2 + R_2 R_3 + R_3 R_1}$$

$$\therefore \quad I_2{}'' = I_1{}'' + I_3{}'' = \frac{(R_1 + R_3)E_2}{R_1 R_2 + R_2 R_3 + R_3 R_1}$$

上記で求めた電流を重ね合わせると，

$$I_1 = I_1{}' + (-I_1{}'') = \frac{(R_2 + R_3)E_1 - R_3 E_2}{R_1 R_2 + R_2 R_3 + R_3 R_1}$$

$$I_2 = (-I_2{}') + I_2{}'' = \frac{-R_3 E_1 + (R_1 + R_3)E_2}{R_1 R_2 + R_2 R_3 + R_3 R_1}$$

$$I_3 = I_3{}' + I_3{}'' = \frac{R_2 E_1 + R_1 E_2}{R_1 R_2 + R_2 R_3 + R_3 R_1}$$

となる.

2.8 鳳・テブナンの定理

鳳・テブナンの定理とは，複数の抵抗や電源からなる回路を，内部抵抗を含む電圧源に置き換えることができるというものである．

Ho-Thevenin's theorem states that any combination of voltage sources, current sources, and resistances with two terminals is electrically equivalent to a single voltage source in series with a single resistance.

● 鳳・テブナンの定理は，2端子をもつ電圧源と電流源，そして抵抗からなるいかなる組合せも，単一抵抗と直列の単一電圧源と電気的に等価であることを示している．

図2・13(a)に示す回路において，抵抗 R を流れる電流 I を求めることを考える．はじめに抵抗 R の両端の端子を a，b とし，そこから抵抗 R を取り外す．そして，図2・13(b)に示すように開放された a-b 間の電圧 V_{ab} を求める．このように開放された端子間の電圧を開放電圧という．

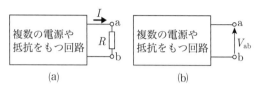

図2・13 鳳・テブナンの定理を適用する回路

次に，図2・14(a)のように回路の電源を取り除いたうえで，a-b間の抵抗 R_{ab} を求める．この場合，電圧源は短絡，電流源は開放にする．開放電圧 V_{ab} と等しい起電力を E_0 とすると，電流 I は次式で表すことができる．

$$I = \frac{E_0}{R_{ab} + R}$$

以上をまとめると，複雑な回路を鳳・テブナンの定理を用いて図2・14(b)に示した一つの等価電圧源と一つの抵抗に置き換えることにより，オームの法則を用いて電流 I を求めることができる．

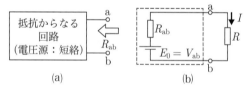

図2・14 鳳・テブナンの定理で等価変換した回路

In Fig.2・15, let us find the voltage V_L across the R_L, and its current I_L.

In order to use Ho-Thevenin's theorem, remove R_L temporarily from the actual circuit. The two open ends then become terminals a and b.

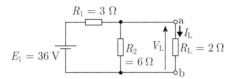

図 2・15　Circuit for example.

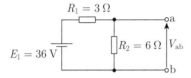

図 2・16　The equivalent circuit.

As shown in Fig.2・16, R_1 and R_2 form a series voltage divider, without R_L.

Furthermore, the voltage V_{R2} across R_2 is the same as the open-circuit voltage across terminals a and b. Therefore, V_{R2} is V_{ab}. This is the E_0 we need for the Ho-Thevenin's equivalent circuit. Using the voltage divider formula,

$$V_{R2} = \frac{6}{3+6} \times 36 = 24 \text{ V}$$

$$V_{ab} = V_{R2} = E_0 = 24 \text{ V}$$

In case of finding R_{ab}, the R_L remains still disconnected. However, the source E_1 is short-circuited. So the circuit looks like Fig.2・17.

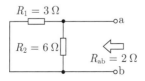

図 2・17　The equivalent circuit.

The R_1 is now in parallel with the R_2. Therefore, this combined resistance R_{ab} is

$$R_{ab} = \frac{R_1 \times R_2}{R_1 + R_2} = \frac{3 \times 6}{3 + 6} = 2\,\Omega$$

In order to find V_L and I_L, we can reconnect R_L to terminals a and b of the Ho-Thevenin's equivalent circuit, as shown in Fig.2・18. Using the voltage divider formula for the R_{ab} and R_L, $V_L = 12\,V$. To find I_L as V_L / R_L, the value is 12 V/2 Ω, which is equal to 6 A.

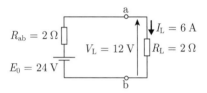

図 2・18　Ho-Thevenin's equivalent circuit.

● 図2・15において，R_Lの両端の電圧 V_L と，その抵抗を流れる電流 I_L を求めてみよう．

鳳・テブナンの定理を用いるために，実際の回路から一時的に R_L を取り除きなさい．そこで，二つの開放端は端子aと端子bになる．

図2・16に示すように，R_L が取り除かれると，R_1 と R_2 は直列の分圧回路を形成する．

さらに，R_2の両端の電圧 V_{R2} は，端子aと端子bの開放電圧に等しい．それゆえ，$V_{R2} = V_{ab}$ となる．これは鳳・テブナンの等価回路で必要な E_0 のことである．分圧回路の公式を用いると，

$$V_{R2} = \frac{6}{3+6} \times 36 = 24 \text{ V}$$
$$V_{ab} = V_{R2} = E_0 = 24 \text{ V}$$

R_{ab}を求めるために，R_L は切り離したままにしておく．しかし，電源 E_1 は短絡する．その結果，回路は図2・17のように表される．R_1 と R_2 は，今や並列の関係にある．それゆえ合成抵抗 R_{ab} は，

$$R_{ab} = \frac{R_1 \times R_2}{R_1 + R_2} = \frac{3 \times 6}{3 + 6} = 2\ \Omega$$

となる.

V_L と I_L を求めるため, 図2・18に示すように, 鳳・テブナンの等価回路の端子aとbの間に R_L を再び接続する. R_{ab} と R_L に分圧回路の公式を用いると, V_L は12 V となる. I_L を V_L/R_L から求めると, その値は12 V/2 Ω = 6 A となる.

2.9　ブリッジ回路

ブリッジ回路は, 周囲を囲む4本の抵抗にある関係が成り立つと, 中央の抵抗を電流が流れなくなる平衡状態となり, これを利用して未知抵抗の測定などが可能になる.

A Wheatstone bridge is used to find the unknown resistance very accurately. A typical Wheatstone bridge is shown in Fig.2·19.

There are two known resistances R_1 and R_2, one variable resistance R_v, and one unknown resistance R_x. If $R_x R_2 = R_1 R_v$, then $V_{CD} = 0$ V and current through galvanometer G = 0 A. By adjusting R_v the current through G is made zero. With zero current in G, the Wheatstone bridge is said to be balanced.

When the Wheatstone bridge is balanced, the value of R_x is obtained as follows.

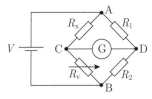

図2・19　Wheatstone bridge.

$$R_\mathrm{x} = \frac{R_1}{R_2} R_\mathrm{v} \tag{2-8}$$

This bridge is used where small changes in resistance are to be measured like in sensor applications.

● ホイートストンブリッジは，未知抵抗値を非常に正確に求めるのに用いられる．代表的なホイートストンブリッジを図2・19に示す．

二つの既知抵抗 R_1 と R_2，一つの可変抵抗 R_v，そして一つの未知抵抗 R_x がある．もし，$R_\mathrm{x}R_2 = R_1R_\mathrm{v}$ ならば，$V_\mathrm{CD} = 0$ V となり，検流計を流れる電流は0アンペアである．R_v を調整し，検流計を流れる電流をゼロにする．検流計を流れる電流がゼロになると，ホイートストンブリッジは平衡状態にあるといわれる．

ホイートストンブリッジが平衡状態のとき，R_x の値は次のように得られる．

$$R_\mathrm{x} = \frac{R_1}{R_2} R_\mathrm{v} \tag{2-8}$$

このブリッジは，センサ応用のような抵抗の微少な変化を測定する場合に用いられる．

Example In Fig.2·20, assume that the bridge is balanced when $R_1 = 100\ \Omega$, $R_2 = 10\ \Omega$, and $R_\mathrm{v} = 30\ \Omega$.

Solve for

a. the value of unknown resistor R_x.

b. the voltages V_CB and V_DB.

c. the total current I, flowing to and from the

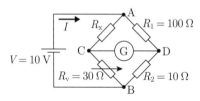

図2・20　Example of Wheatstone bridge.

voltage source V.

例題 図2・20において，$R_1 = 100\ \Omega$，$R_2 = 10\ \Omega$，$R_v = 30\ \Omega$のとき，ブリッジは平衡状態にあると仮定しなさい．

a. 未知抵抗R_xの値を求めよ．

b. 電圧V_{CB}とV_{DB}を求めよ．

c. 電圧源Vとの間で流れる全電流Iを求めよ．

Solution（答） a. Using Formula(2-8), R_x is calculated as follows.

a. (2-8)式を用いて，R_xは以下のように計算される．

$$R_x = \frac{R_1}{R_2}\, R_v = \frac{100}{10} \times 30 = 300\ \Omega$$

b. As the bridge is balanced, $V_{AC} = V_{AD}$ and $V_{CB} = V_{DB}$.

Therefore,

b. ブリッジは平衡状態なので，V_{AC}とV_{AD}は等しく，V_{CB}とV_{DB}も等しい．

それゆえ，

$$V_{CB} = \frac{R_v}{R_x + R_v}\, V = \frac{30}{300 + 30} \times 10 = 0.91\ \mathrm{V}$$

$$V_{DB} = \frac{R_2}{R_1 + R_2}\, V = \frac{10}{100 + 10} \times 10 = 0.91\ \mathrm{V}$$

c. When the bridge is balanced, combined resistance R_{AB} is obtained as follows.

c. ブリッジが平衡状態にあるとき，合成抵抗R_{AB}は以下のように得られる．

$$R_{AB} = \frac{(R_1 + R_2) \times (R_x + R_v)}{(R_1 + R_2) + (R_x + R_v)} = \frac{110 \times 330}{110 + 330}$$
$$= 82.5\ \Omega$$

$$\therefore \quad I = \frac{V}{R_{AB}} = \frac{10}{82.5} = 121 \,\text{mA}$$

The Murray loop method applying the Wheatstone bridge is used as a method of finding out the fault location when a fault occurs in an underground cable.

● ホイートストンブリッジを応用したマーレーループ法は，地中送電線に故障が発生した場合に故障箇所を見つける方法として用いられる．

2.10　電力と電力量

電流によって生じる1秒間当たりの電気エネルギーを電力といい，電力をある時間抵抗に加えたときの電気エネルギーの総量を電力量という．

The unit of electric power is the watt (W), named after James Watt (1736-1819). One watt of power equals the work done in one second by one volt of potential difference in moving one coulomb of charge.

Remember that one coulomb per second is an ampere. Therefore, power in watts equals the product of volts times amperes.

$$P = VI$$

● 電力の単位であるワット（W）は，ジェームズ・ワット（1736-1819）にちなんで命名された．1ワットの電力は，1クーロンの電荷が1ボルトの電位差のある所を1秒間に移動する仕事に等しい．

1秒当たり1クーロンの電荷の移動が1アンペアであることを思い出してほしい．それゆえ，ワットを単位とする電力は電圧と電流の積に等しい．

電力 ＝ 電圧 × 電流

When a current flows through a resistance, heat is produced because friction between the moving free electrons and the atoms obstructs the path of electron flow. The heat is evidence that power is used in producing current. The power is generated by the source of applied voltage and consumed in the resistor as heat.

Since power is dissipated in the resistance of a circuit, it is convenient to express the power in terms of the resistance R.

$$P = VI = I^2R = \frac{V^2}{R} \tag{2-9}$$

where V is the voltage across the resistance in Volts, I the current through the resistance in Amperes, and R the resistance of the resistor in Ohm's.

The power P in watts(W) is equal to the energy E in joules(J), divided by the time period t in seconds(s). The joule is a general unit of work or energy.

To summarize these definitions,

　　　1 J = 1 W·s

or

　　　1 W = 1 J/s

● 抵抗を電流が流れると，移動する自由電子と原子との間の摩擦が電子の流れを妨げるため，熱が発生する．熱は，電流をつくり出すうえで電力が使われるという証拠になる．その電力は印加電源によって生成され，抵抗で熱として消費される．

電力は回路内の抵抗で消費されるので，抵抗 R を用いて表すと便利である．

$$P = VI = I^2R = \frac{V^2}{R} \tag{2-9}$$

ここで，V は抵抗の両端の電圧をボルトで，I は抵抗を流れる電流をアンペアで，R は抵抗値をオームでそれぞれ表す．

ワット（W）単位の電力 P は，ジュール（J）単位のエネルギー E を秒

2 電気回路の基礎

（s）単位の時間間隔 t で割ったものに等しい．ジュールは，仕事やエネルギーの一般的な単位である．

これらの定義を要約すると，

$$1 \text{ J} = 1 \text{ W·s}$$

または，

$$1 \text{ W} = 1 \text{ J/s}$$

Exampie1 Calculate the power dissipated in a 10 kΩ resistor with a 5 mA current through the resistor.

例題1 10 kΩ の抵抗に 5 mA の電流を流したときに消費される電力を計算しなさい．

Solution（答） Using Eq.(2-9), P is calculated as follows.

（2-9）式を用いて，P は以下のように計算される．

$$P = I^2 R = (5 \times 10^{-3})^2 \times 10 \times 10^3$$
$$= 250 \text{ mW} = 0.25 \text{ W}$$

Electric energy is defined as the ability to do work. Electric energy W is equal to electric power P times time t.

$$W = Pt = VIt = I^2 Rt = \frac{V^2}{R}t \qquad (2\text{-}10)$$

In the equation (2-10), V is in volts, I in amperes and R in Ohm's. If t is expressed in hours, W will be in watt-hours. If t is expressed in seconds, W will be in watt-seconds or joules.

● 電力量は，仕事をする能力として定義される．電力量 W は電力 P と時間 t を掛けたものに等しい．

$$W = Pt = VIt = I^2 Rt = \frac{V^2}{R}t \qquad (2\text{-}10)$$

（2-10）式において，V の単位はボルト，I の単位はアンペア，そして R の単位はオームである．t が時間単位で表されていれば，W はワット時で表される．t が秒単位で表されていれば，W はワット秒またはジュールと

なる.

Exampie2 How much energy(J) is supplied to a 100 Ω resistor which is connected to a 150 V supply for 1 hour?

例題2 150 V電源に1時間接続された100 Ωの抵抗に, どれだけのエネルギー (J) が供給されるか.

Solution (答) Using equation (2-9), P is calculated as follows.
(2-9) 式を用いて, P は以下のように計算される.

$$P = \frac{V^2}{R} = \frac{150^2}{100} = 225 \text{ W}$$

Using equation (2-10), W is calculated as follows.
(2-10) 式を用いて, W は以下のように計算される.

$$W = Pt = 225 \times 1 \times 60 \times 60 = 810\,000 \text{ J}$$
$$= 810 \text{ kJ}$$

2.11 最大電力供給の定理

電源に内部抵抗がある場合には, 電源の内部抵抗と負荷抵抗が等しいときに回路に最大の電力が供給される. これが最大電力供給の定理である.

The maximum power transfer theorem states that in a linear, bilateral direct-current network, the maximum power is delivered to the load when the load resistance is equal to the internal resistance of the source.

Let us consider a power source modelled by a Ho-Thevenin's equivalent circuit with a load resistance R_L as shown in Fig.2·21. The source resistance is r and the open-circuit voltage of the source is E.

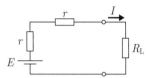

図2・21 Maximum power transfer theorem.

● 最大電力供給の定理は，線形で両方向性直流回路網において，負荷抵抗が電源の内部抵抗に等しいときに負荷に最大電力が供給されることを示している．

図2・21に示すような，負荷抵抗 R_L を接続した鳳・テブナンの等価回路でモデル化された電源を考えてみよう．電源の抵抗は r，電源の開放電圧は E である．

In Fig.2·21, the current in this circuit is calculated using Ohm's law.

$$I = \frac{E}{r + R_L} \, [\text{A}]$$

The power dissipated in the load resistance, P_L is as follows.

$$P_L = I^2 R_L = \left(\frac{E}{r + R_L}\right)^2 R_L = \frac{R_L}{(r + R_L)^2} E^2 \, [\text{W}]$$

In the above equation, R_L is variable. Therefore, the condition for maximum power delivered to the load is determined by differentiating load power with respect to the load resistance and equating it to zero.

$$\frac{dP_L}{dR_L} = \frac{\left\{(R_L + r)^2 - 2R_L(r + R_L)\right\}}{(R_L + r)^4} E^2$$

$$= \frac{r - R_L}{(R_L + r)^3} E^2 = 0$$

$$\rightarrow R_L = r$$

This is the condition for maximum power transfer, which states that power delivered to the load is maximum, when the load resistance R_L matches with Thevenin's resistance r of the network.

Under this condition, power transfer to the load is

$$P_L = \left(\frac{E}{r+r}\right)^2 r = \frac{E^2}{4r} \text{ [W]}$$

The power P_L is called available power.

Total power P_T transferred from the source is as follows.

$$P_T = I^2(r + R_L) = 2I^2r = 2I^2R_L$$

● 図2・21において，この回路を流れる電流は，オームの法則を用いて計算される.

$$I = \frac{E}{r + R_L} \text{ [A]}$$

負荷抵抗で消費される電力 P_L は，以下のとおりである.

$$P_L = I^2 R_L = \left(\frac{E}{r+R_L}\right)^2 R_L = \frac{R_L}{(r+R_L)^2} E^2 \text{ [W]}$$

上記の式において，R_L は変数である．それゆえ，負荷に供給される最大電力の条件は，その電力を負荷抵抗で微分し，ゼロに等しいと考えることで決定される.

$$\frac{\mathrm{d}P_L}{\mathrm{d}R_L} = \frac{\left\{(R_L + r)^2 - 2R_L(r + R_L)\right\}}{(R_L + r)^4} E^2 = \frac{r - R_L}{(R_L + r)^3} E^2 = 0$$

$$\rightarrow \quad R_L = r$$

これは最大電力供給の条件で，回路網において負荷抵抗 R_L が内部抵抗 r と一致するとき，負荷に供給される電力は最大となることを表している.

この条件で負荷に供給される電力は

$$P_L = \left(\frac{E}{r+r}\right)^2 r = \frac{E^2}{4r} \text{ [W]}$$

この電力は有能電力と呼ばれる.

電源から供給される全電力 P_T は，以下のとおりである.

$$P_T = I^2(r + R_L) = 2I^2r = 2I^2R_L$$

③ 電気回路における物理現象

3.1 *RC* 直列回路の過渡現象

　回路のスイッチを閉じたとき，電流や電圧などがある定常状態から別の定常状態になるまでの変動現象を過渡現象（transient phenomena）という．*RC* 直列回路に電圧を印加した場合，コンデンサにはすぐに電圧がかからないで徐々に充電されていく．

A capacitor stores energy in the form of electric charge. Fig.3・1 shows an RC series circuit consisting of a resistance of resistance R, a capacitor of capacitance C in series connected to a voltage source E through a switch.

● 　コンデンサはエネルギーを電荷の形で蓄える．図3・1は抵抗 R の抵抗器，スイッチを介して電圧源 E と直列接続された静電容量 C のコンデンサからなる RC 直列回路を表す．

In Fig.3・1, let us assume that the capacitor is initially uncharged. When the switch is closed, the capacitor will gradually charge up until the voltage across it reaches the source voltage.

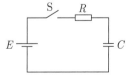

図 3・1　RC series circuit.

The voltage V_R across the resistance and the voltage V_C across the capacitor are as follows using the current i flowing in the circuit:

$$V_R = iR$$

$$V_C = \frac{1}{C} \int_0^t i \, \mathrm{d}t$$

By applying Kirchhoff's Voltage Law, the following equation is obtained:

$$iR + \frac{1}{C} \int_0^t i \, \mathrm{d}t = E$$

$q = \int_0^t i \, \mathrm{d}t$, $i = \dfrac{\mathrm{d}q}{\mathrm{d}t}$, so the above equation can be written as follows.

$$R \frac{\mathrm{d}q}{\mathrm{d}t} + \frac{q}{C} = E$$

Separation of variables may be used to solve the above differential equation.

$$\frac{1}{EC - q} \mathrm{d}q = \frac{1}{RC} \mathrm{d}t$$

Integrate both sides of the equation separately.

$$\int \frac{1}{EC - q} \mathrm{d}q = \int \frac{1}{RC} \mathrm{d}t$$

$$\log(EC - q) = -\frac{1}{RC} t + K$$

$$EC - q = \mathrm{e}^{-\frac{t}{RC}} \mathrm{e}^K$$

where K is a constant of integration, which comes from the initial conditions: $q = 0$ at $t = 0$.

$$\therefore \quad q(t) = EC\left(1 - e^{-\frac{t}{RC}}\right)$$

$$i(t) = \frac{\mathrm{d}q(t)}{\mathrm{d}t} = \frac{E}{R}e^{-\frac{t}{RC}}$$

And finally the step response of the capacitor voltage can be expressed by

$$V_C(t) = \frac{q(t)}{C} = E\left(1 - e^{-\frac{t}{RC}}\right) = E\left(1 - e^{-\frac{t}{\tau}}\right)$$

This response is shown in Fig.3·2, where the vertical axis shows the voltage of the capacitor, $V_C(t)$.

In the above equation, $\tau = RC$ is called time constant of the RC series circuit. The time constant is defined as the time when the voltage $V_C(t)$ reaches approximately 63 % of the source voltage E.

● 図3・1において，コンデンサは初期状態として充電されていないものとする．スイッチを閉じると，コンデンサは両端の電圧が電源電圧に達するまで徐々に充電される．

抵抗の両端の電圧 V_R およびコンデンサの両端の電圧 V_C は，回路に流れる電流 i を用いて以下のようになる：

$$V_R = iR$$

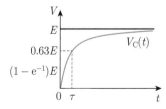

図3・2　Variation of voltage with time in RC series circuit.

$$V_\mathrm{C} = \frac{1}{C} \int_0^t i\, \mathrm{d}t$$

キルヒホッフの電圧則を適用すると，以下の式が得られる．

$$iR + \frac{1}{C} \int_0^t i\, \mathrm{d}t = E$$

$q = \int_0^t i\, \mathrm{d}t$，$i = \dfrac{\mathrm{d}q}{\mathrm{d}t}$ であるので，上記の式は以下のように書くことができる．

$$R\frac{\mathrm{d}q}{\mathrm{d}t} + \frac{q}{C} = E$$

上記の微分方程式を解くのに，変数分離を用いることができる．

$$\frac{1}{EC - q}\, \mathrm{d}q = \frac{1}{RC}\, \mathrm{d}t$$

方程式の両辺を，それぞれ積分しなさい．

$$\int \frac{1}{EC - q}\, \mathrm{d}q = \int \frac{1}{RC}\, \mathrm{d}t$$

$$\log(EC - q) = -\frac{1}{RC}\, t + K$$

$$EC - q = \mathrm{e}^{-\frac{t}{RC}}\, \mathrm{e}^{K}$$

ここで，K は積分定数で，初期条件：$t = 0$ で $q = 0$ から得られる．

$$\therefore \quad q(t) = EC\left(1 - \mathrm{e}^{-\frac{t}{RC}}\right)$$

$$i(t) = \frac{\mathrm{d}q(t)}{\mathrm{d}t} = \frac{E}{R}\, \mathrm{e}^{-\frac{t}{RC}}$$

最後に，コンデンサの両端の電圧のステップ応答は，

$$V_\mathrm{C}(t) = \frac{q(t)}{C} = E\left(1 - \mathrm{e}^{-\frac{t}{RC}}\right) = E\left(1 - \mathrm{e}^{-\frac{t}{\tau}}\right)$$

で表される．

この応答は図3・2で示される．ここで，縦軸はコンデンサの電圧，$V_\mathrm{C}(t)$ を表す．

上式で，$\tau = RC$ は RC 直列回路の時定数と呼ばれる．時定数は，電圧 $V_\mathrm{C}(t)$ の値が電源電圧 E の約63 % に達する時間として定義される．

3・2 磁気現象と磁極に働く力

　磁石は，鉄を引き付ける性質をもっている．このような性質を磁性といい，磁性のもとになるものを磁気という．磁石はN極とS極の二つの磁極からできている．N極とS極は引き合い，N極とN極，S極とS極は反発する．方位磁針は，ほぼ南北を指して静止する．これは，地球は大きな磁石になっていて，地理学上の北極付近にS極，南極付近にN極があると考えられる．このように地球に存在する磁気を地磁気という．

　A magnet attracts magnetic material with a force called magnetism. For example, iron is attracted by a magnet with the force of magnetism.

　A bar magnet is a rectangular object of uniform cross-section that attracts pieces of ferrous objects. Both ends of the magnet are called magnetic poles. The magnet is made up of two magnetic poles, north pole and south pole. The north pole will attract the south pole of another magnet, and repel the north pole. These forces are magnetism. The space surrounding a magnet, in which magnetism is exerted, is called a magnetic field.

　Although the magnetic field cannot be seen, evidence of its force can be seen when fine iron filings are sprinkled on a sheet of paper placed over a bar magnet.

　To describe the phenomena related to magnets, lines are used to depict the force existing in the area surrounding the magnet. These lines are called lines of magnetic force. These lines do not exist actually, but are imaginary lines that are used to illustrate and

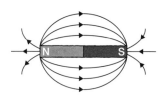

図 3・3 Line of magnetic force.

describe the pattern of the magnetic field.

As shown in Fig.3·3, lines of magnetic force are defined to originate from the north pole of the magnet, then pass through the surrounding space and go into the south sole. Lines of magnetic force are closed curves. Outside the magnet their direction is from the north pole to the south pole and inside the magnet these are from south pole to north pole.

Since the magnetic field intensity depends on the number of lines of magnetic force, the magnetic field is strongest at both magnetic poles of the magnet.

● 磁石は，磁力と呼ばれる力で磁性材料を引き付ける．例えば，鉄は磁気の力で磁石に引き付けられる．

棒磁石は，鉄製の物体を引き付ける一様断面をもつ長方形の棒である．その磁石の両端を磁極という．磁石はN極とS極の二つの磁極からできている．N極はほかの磁石のS極を引き付け，N極とは反発する．これらの力が磁力である．磁力が作用する磁石の周りの空間を磁界と呼ぶ．

磁界は目に見えないが，棒磁石の上に置かれた紙に細かい鉄粉を振りかけると，その力の存在が確認できる．

磁石に関連する現象を説明するために，磁石の周りに存在する力を表すのに線が用いられる．これらの線は磁力線と呼ばれる．これらの線は実際には存在しないが，磁界のパターンを説明するために用いられる仮想線である．

図3・3に示すように，磁力線は磁石のN極から始まり，周囲の空間を

通過してからS極に入ると定義されている．磁力線は閉曲線である．磁石
の外側では，それらの方向はN極からS極に向かい，磁石の内側ではこれ
らはS極からN極に向かう．磁界の強さは磁力線の数に依存するため，磁
界は磁石の両磁極で最も強い．

When a bar magnet is brought close to an iron nail as shown in
Fig.3・4, it is attracted to the magnet. The nail holds its magnetism
as long as the magnet is held close to it or in contact with it.
The magnetized nail magnetizes several other nails. However,
magnetized nails lose their magnetism when they are separated
from the magnet. Magnetic induction is the process by which iron
piece like iron nail is magnetized by the magnetic field. During
magnetic induction, the closer end of the magnetic material has
unlike polarity and the more distant end has like polarity.

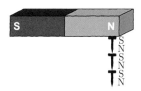

図3・4 Magnetic induction.

● 図3・4に示すように，鉄くぎの近くに棒磁石を近づけると，鉄くぎは
磁石に引き付けられる．くぎは磁石がそれに近づくか接触している限り，
その磁気を保持する．磁化されたくぎは，いくつかのほかのくぎを磁化す
る．しかし，磁化されたくぎは，磁石から離れると磁気を失う．磁気誘
導とは，鉄くぎのような鉄片が磁界によって磁化される過程である．磁気
誘導中，磁性材料のより近い端部は異なる極性を有し，より遠い端部は同
じ極性を有する．

Let us assume that a bar magnet is suspended from its center by
a string. If the north pole of another bar magnet is brought near

the north pole of the suspended bar magnet, there will be a force of repulsion between these poles.

On the other hand, there will be a force of attraction when the north pole of one magnet approaches the south pole of another one.

● 棒磁石がその中心から糸でつり下げられていると仮定する．別の棒磁石のN極をつり下げられた棒磁石のN極に近づけると，これらの極の間に反発力が生じる．

一方，磁石のN極を別の磁石のS極に近づけると，吸引力が働く．

The force (F) in newton between the magnetic poles of two magnets is directly proportional to the product of the strength of their poles and inversely proportional to the square of the distance between them.

Coulomb's law in magnetism is stated as follows.

$$F = \frac{1}{4\pi\mu}\frac{m_1 m_2}{r^2} \text{ [N]}$$

where r = distance between the poles in meter. $\mu = \mu_0 \mu_r$, μ_r = relative permeability and μ_0 = absolute permeability of vacuum. $\mu_0 = 4\pi \times 10^{-7}$ H/m and for air or vacuum $\mu_r = 1$. Moreover, m_1

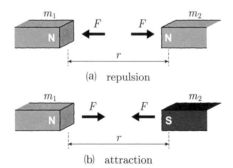

(a) repulsion

(b) attraction

図3・5 Force between two magnetic poles.

and m_2 are the magnetic intensities of the poles and expressed in the unit weber (Wb). If m_1 and m_2 are of the same sign, F is positive and the force is repulsive. If m_1 and m_2 are of opposite signs, F is negative and the force is attractive. See Fig.3・5.

● 二つの磁石の磁極間に働くニュートンを単位とする力 (F) は，両磁極の強さの積に比例し，両磁極間の距離の2乗に反比例する．

磁気に関するクーロンの法則は次のように述べられる．

$$F = \frac{1}{4\pi\mu} \frac{m_1 m_2}{r^2} [\text{N}]$$

ここで，r はメートルを単位とする2極間の距離である．μ は $\mu_0\mu_r$ に等しく，μ_r は比透磁率，μ_0 は真空透磁率である．$\mu_0 = 4\pi \times 10^{-7}$ H/m で，空気中または真空中では $\mu_r = 1$ である．さらに，m_1 と m_2 は磁極の強さで，単位はウェーバ (Wb) で表される．m_1 と m_2 が同符号の場合は，F の大きさは正となって反発力を表し，m_1 と m_2 が異符号の場合は，F の大きさは負となって吸引力を表す．図3・5を参照してください．

Example A magnetic pole having force of 0.5×10^{-3} Wb is placed in a magnetic field at a distance of 25 cm from another magnetic pole in vacuum. It is experiencing a force of 0.5 N. Determine the force of the other magnetic pole and the direction of the force acting between the two magnetic poles.

例題 真空中で磁極の強さ 0.5×10^{-3} Wb の磁極が，同極の他の磁極から25 cm離れた磁界に置かれている．この磁極は0.5 Nの力を受けている．もう一方の磁極の強さと二つの磁極間に作用する力の向きを求めなさい．

Solution (答) According to Coulomb's law,

クーロンの法則より，

$$F = \frac{1}{4\pi\mu_0} \frac{m_1 m_2}{r^2} = 6.33 \times 10^4 \times \frac{m_1 m_2}{r^2} [\text{N}]$$

$$\therefore \ m_2 = \frac{r^2}{6.33 \times 10^4} \times \frac{F}{m_1}$$

$$= \frac{0.25^2}{6.33 \times 10^4} \times \frac{0.5}{0.5 \times 10^{-3}}$$

$$= 9.87 \times 10^{-4} \ \mathrm{Wb}$$

$$F = 9.87 \times 10^{-4} \ \mathrm{Wb} \ (\mathrm{Repulsion}) \quad (反発)$$

When a magnetic pole of strength m [Wb] is brought into a magnetic field of intensity H [A/m], the magnetic force F [N] acting on the magnetic pole is expressed as follows.

$$F = mH$$

As already mentioned, the magnetic field can be visualized as lines of magnetic force. The intensity of magnetic field corresponds to the density of lines of magnetic force. Magnetic flux is a measure of the number of magnetic field lines passing through an area. The quantity symbol we use for the magnetic flux is the Greek letter ϕ and the unit is Wb.

The magnetic flux density is the amount of magnetic flux per unit area of a section perpendicular to the direction of magnetic flux. The magnetic flux density B is given by the following formula.

$$B = \frac{\phi}{S}$$

where ϕ is the flux through an area S. The quantity symbol of magnetic flux density is B, and the unit is Tesla [T].

● 磁界の強さ H [A/m] の磁界中に，m [Wb] の磁極を置いたとき，この磁極に作用する磁力 F [N] は次のように表される．

$$F = mH$$

すでに述べたように，磁界は磁力線として視覚化できる．磁界の強さは，磁力線の密度に対応する．磁束は，ある領域を通過する磁力線の数の尺度である．磁束に用いる量記号はギリシャ文字の ϕ，単位は Wb である．

磁束密度は，磁束の向きに垂直な断面の単位面積当たりの磁束の量である．磁束密度 B は以下の式で与えられる．

$$B = \frac{\phi}{S}$$

ここで，ϕ は面積 S を貫く磁束である．磁束密度の量記号は B，単位はテスラ [T] である．

3.3 アンペアの右ねじの法則

直線導体に電流が流れると，その付近に磁界が生じる．この磁界の向きはアンペアの右ねじの法則によって知ることができる．

The direction of the magnetic field can be determined by Ampere's right-handed screw rule. In Fig.3・6, if a screw points in the direction of the current, the direction of the lines of magnetic force is shown by the direction your hand must move to turn the screw.

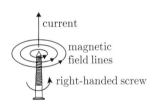

図 3・6 Ampere's right-handed screw rule.

● 磁界の向きは，アンペアの右ねじの法則によって決まる．図3・6において，電流の向きをねじの進む向きにとると，磁力線の向きは，ねじを回すために手を動かす向きで示される．

As shown in Fig.3・7, a helical coil of wire with current passing through each loop of the coil is called a solenoid. By wrapping

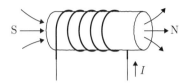

図3・7　Magnetic field of a solenoid.

the wire many times around the cylinder, the magnetic field due to the wires can become quite strong. More loops will bring about a stronger magnetic field.

● 図3・7に示すように，コイルの各ループを電流が流れるらせん状のコイルは，ソレノイドと呼ばれる．円筒の周りに導線を数多く巻くことによって，磁界が非常に強くなる．巻数が多いほど，より強い磁界をもたらす．

In Fig. 3·7, if you assume your right hand gripping the solenoid so that your fingers follow the direction of the current then your outstretched thumb points towards the north pole. When a current is passed through the solenoid, a magnet is formed because the magnetic field is concentrated inside the coil.

By adding an iron core within the solenoid, the iron core becomes magnetized. This produces a powerful electromagnet.

The existence of the ferromagnetic material within the soleniod greatly increases the strength of the magnetic field.

Electromagnet is particularly useful because it can be switched on and off, and strengthened by increasing the current flowing through it.

● 図3・7において，右手がソレノイドを握って指を電流の向きに合わせると，広げた親指はN極の方を指す．ソレノイドに電流を流すと，コイル内に磁界が集中して磁石が形成される．

　ソレノイド内に鉄心を追加することにより，鉄心は磁化される．これは強力な電磁石をつくり出す．

　　ソレノイド内に強磁性体が存在すると，磁界の強さは増大する．

　　電磁石は，オンとオフを切り替えることができ，それを流れる電流を増加させることによって強くすることができるので，特に有用である．

3.4 フレミングの左手の法則

　フレミングの左手の法則は，磁界中に導体を置き，導体に電流を流したときの電磁力についての法則である．フレミングの左手の法則は，電流の向きと磁界の向きから導体の動く力の向きを求めるときに用い，モータの原理を指し示す法則である．

According to Fleming's left-hand rule, if the thumb, index finger and middle finger of the left hand are stretched to be perpendicular to each other as shown in Fig.3・8, and if the index finger represents the direction of magnetic flux density and the middle finger represents the direction of current, then the thumb will point the direction of force.

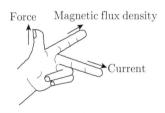

図 3・8　Fleming's left-hand rule.

● 　フレミングの左手の法則によれば，図3・8に示すように左手の親指，人差し指，中指を互いに直角になるように開いて，人差し指が磁束密度の向き，中指が電流の向きを表すならば，親指は力の向きを指す．

Electric motor works based on Fleming's left-hand rule. An electric motor simply uses a current and a magnetic field to create motion as the interaction of the magnetic field and electric current exerts force. This means that electric energy is converted into kinetic energy.

● モータは，フレミングの左手の法則に基づいて動作する．磁界と電流との相互作用が力を及ぼすので，モータは単に電流と磁界を利用して運動をつくり出している．すなわち，電気エネルギーが運動エネルギーに変換される．

When a conductor carrying an electric current is placed in a magnetic flux density, the conductor undergoes a force which is perpendicular to both the directions of the magnetic flux density and the electric current. In Fig.3·9, the force F is proportional to the magnetic flux density B, the electric current I on the conductor and the length l of the conductor that is in the magnetic field.

$$F = BIl \, [\text{N}]$$

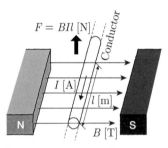

図3・9　Force on a conductor in a magnetic field.

● 電流が流れている導体を磁束密度の中に置くと，導体は，磁束密度と電流の両方の向きに垂直な力を受ける．図3・9において，力 F は，磁束密度 B，導体を流れる電流 I，および磁界中の導体の長さ l に比例する．

$$F = BIl \, [\text{N}]$$

Exercise　Determine the current required in a 400 mm length of conductor of an electric motor, when the conductor is situated at right-angles to a magnetic field of flux density 1.2 T, if a force of 1.92 N is to be exerted on the conductor.

例題 導体が磁束密度1.2 Tの磁界に対して直角に置かれているとき，導体に1.92 Nの力を働かせるためには，モータの長さ400 mmの導体に必要な電流を求めなさい．

Solution (答) $F = 1.92$ N, $l = 400$ mm $= 0.4$ m and $B = 1.2$ T.

Since $F = BIl$, then $I = F/Bl$

$F = 1.92$ N, $l = 400$ mm $= 0.4$ mそして$B = 1.2$ T.

$F = BIl$より，$I = F/Bl$

$$I = \frac{1.92}{1.2 \times 0.4} = 4 \text{ A}$$

3.5 フレミングの右手の法則

フレミングの右手の法則は，磁界中にある導体を動かしたときに，導体に生じる起電力についての法則である．フレミングの右手の法則は，発電機の原理を指し示す法則である．

Let the thumb, index finger and middle finger of the right hand stretch to be mutually perpendicular, as shown in Fig.3・10. According to Fleming's right-hand rule, if both the thumb is in the

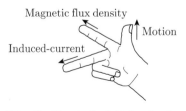

Magnetic flux density

Motion

Induced-current

図3・10 Fleming's right-hand rule.

direction of motion and the index finger is in the direction of magnetic flux density, the middle finger will give the direction of induced-current.

● 図3・10に示すように，親指，人差し指，中指を開いて互いに直角になるようにする．フレミングの右手の法則によれば，親指が運動の向き，人差し指が磁束密度の向きにある場合，中指が誘導電流の向きを指す．

Fleming's right-hand rule applies to generators that generate current by motion.

In Fig.3・11, if a conductor is moving at right angle to the magnetic flux density, the direction of the induced current can be found using Fleming's right-hand rule. When the thumb, index finger and middle finger of the right hand are perpendicular to each other, the middle finger gives the direction of induced current if the index finger points the direction of magnetic flux density and the thumb points the direction of motion.

● フレミングの右手の法則は，運動により電流を発生させる発電機に適用される．

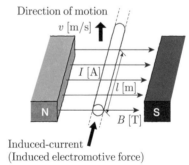

図3・11 Fleming's right-hand rule of generated induced electromotive force.

　図3・11において，導体が磁束密度に対して垂直に運動する場合，誘導電流の方向はフレミングの右手の法則を用いて求めることができる．右手の親指，人差し指，中指が互いに直角の場合，人差し指が磁束密度の向き，親指が運動の向きを指すならば，中指は誘導電流の向きを与える．

The general equation for the voltage induced into a conductor that is cutting line of magnetic force at angle θ is given by:

$$e = Blv \sin \theta$$

where:

e = induced electromotive force [V]

B = magnetic flux density [T]

l = length of conductor [m]

v = velocity of conductor [m/s]

θ = angle between the direction of flux density and conductor velocity [°]

The direction of the induced electromotive force can be deduced by applying Fleming's right-hand rule.

● 角度θで磁力線を横切る導体に誘導される電圧の一般式は，
$$e = Blv \sin \theta$$
で与えられる．

　ここで，eは誘導起電力 [V]，Bは磁束密度 [T]，lは導体の長さ [m]，vは導体の速度 [m/s]，θは磁束密度と導体の速度とのなす角度 [°] である．

　誘導起電力の向きは，フレミングの右手の法則を当てはめることで導かれる．

Example　A conductor 300 mm long moves at a uniform speed of 4 m/s at right-angles to a uniform magnetic field of flux density 2.5 T. Determine the current flowing in the conductor when (a) its end are open-circuited, (b) its ends are connected to 20 Ω load resistance.

例題 長さ300 mmの導体が磁束密度2.5 Tの一様な磁界と直角に4 m/sの一定速度で移動する．(a)その端部が開放されているとき，(b)その端部が20 Ωの負荷抵抗に接続されているときに導体を流れる電流を求めよ．

Solution（答） When a conductor moves in a magnetic field it will have an induced electromotive force in it, but this electromotive force can only produce a current if there is a closed circuit.

導体が磁界中を運動すると，導体に誘導起電力が生じるが，回路が閉じている場合に限りこの起電力により電流が流れる．

Induced electromotive force is calculated as follows.

誘導起電力は，次のように計算される．

$$e = Blv = 2.5 \times \frac{300}{1\,000} \times 4 = 3 \text{ V}$$

(a) If the ends of the conductor are open-circuited, no current will flow even though 3 V has been induced.

導体の端部が開放されていると，3 Vが誘導されても電流は流れない．

(b) From Ohm's law, $I = \dfrac{e}{R} = \dfrac{3}{20} = 0.15$ A or 150 mA

オームの法則から，$I = \dfrac{e}{R} = \dfrac{3}{20} = 0.15$ A または 150 mA．

3.6 ファラデーの電磁誘導

　電磁誘導によってコイルに誘導される起電力は，そのコイルと鎖交する磁束の時間に対する変化の割合に比例する．これをファラデーの電磁誘導の法則という．

Consider a solenoid and a bar magnet, as shown in Fig 3・12. When the bar magnet is moved closer to or away from the solenoid connected with a galvanometer, the needle of the galvanometer shows deflection. From this experiment, it can be seen that when the magnetic flux cutting across turns of wire in the solenoid changes, the electromotive force is induced and current flows in the circuit. This phenomenon is called electromagnetic induction. Also, the electromotive force generated by this is called induced electromotive force, and the current is called induced current.

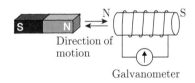

Direction of motion

Galvanometer

図3・12　Electromagnetic induction.

● 図3・12に示すように，ソレノイドと棒磁石を考える．棒磁石を，検流計を接続したソレノイドに近づけたり離したりすると，検流計の針が振れる．この実験から，ソレノイドの巻線を横切る磁束が変化すると，起電力が誘導され，回路に電流が流れることがわかる．この現象は電磁誘導と呼ばれる．さらに，これによって生じる起電力を誘導起電力，流れる電流を誘導電流という．

Faraday's law of electromagnetic induction states that the induced electromotive force is proportional to the rate of change of magnetic flux linked with the coil.

Faraday's law of electromagnetic induction can be written as follows.

$$e = -N \frac{d\phi}{dt}$$

where

e = induced electromotive force [V]

ϕ = magnetic flux [Wb]

N = number of turns of the coil

Unlike Faraday's law of electrolysis, this relationship is called Faraday's law of electromagnetic induction.

The negative sign in Faraday's law of electromagnetic induction shows that the induced electromotive force is in opposition to the flux change. This is known as Lenz's law.

● ファラデーの電磁誘導の法則は，誘導起電力は，コイルと鎖交する磁束の変化率に比例すると述べられる．

　ファラデーの電磁誘導の法則は，以下のように書くことができる．

$$e = -N \frac{d\phi}{dt}$$

ここで，e [V]は誘導起電力，ϕ [Wb]は磁束，Nはコイルの巻数である．

　ファラデーの電気分解の法則とは異なり，この関係はファラデーの電磁誘導の法則と呼ばれている．

　ファラデーの電磁誘導の法則の負の符号は，誘導起電力が磁束の変化を妨げることを示している．これはレンツの法則として知られている．

Example　A circular coil of 200 turns and radius 20 cm is placed in and perpendicular to a magnetic flux density of 2.5×10^{-2} T. If the magnetic flux density is reduced to zero in 0.01 s, what electromotive force is induced in the coil?

例題 磁束密度 2.5×10^{-2} T と垂直に，巻数が 200 回，半径が 20 cm の円形コイルが配置されている．磁束密度を 0.01 s でゼロにすると，コイルに発生する誘導起電力はいくらか．

Solution (答) Magnetic flux through the coil is given as follows.

$$\phi = BS$$

For a circular coil, the area S is given by πr^2.

コイルを鎖交する磁束は，$\phi = BS$ で与えられる．
円形コイルの場合，面積は πr^2 で与えられる．

$$\phi = 2.5 \times 10^{-2} \times \pi \times 0.20^2 = 3.14 \times 10^{-3} \text{ Wb}$$

Hence, the induced electromotive force is obtained from the following equation.

したがって，誘導起電力は以下の式から得られる．

$$e = -N\frac{\mathrm{d}\phi}{\mathrm{d}t} = -200 \times \frac{0 - 3.14 \times 10^{-3}}{0.01}$$

$$= 61.8 \text{ V}$$

3.7 レンツの法則

　電磁誘導によって生じる起電力は，元の磁束の変化を妨げる向きに生じる．これをレンツの法則という．

　Lenz's law is the fundamental principle for determining the direction of an induced voltage or current.

　As the magnet approaches the coil as shown in Fig.3·13(a), the magnetic flux interlinking the coil increases.

　According to Faraday's law of electromagnetic induction, when there is change in flux, an electromotive force and hence current is

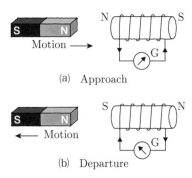

(a) Approach

(b) Departure

図3・13 Demonstration of Lenz's law.

induced in the coil and this current will create its own magnetic field. The galvanometer indicates both the direction and the magnitude of the current.

Lenz's law states that the induced electromotive force generated by electromagnetic induction creates in a direction that opposes the change in the magnetic flux interlinking across the coil. This means that the left end of the coil becomes north pole and the north poles of the magnet and the coil repel each other.

When the magnetic pole of the coil side is known, the direction of the induced current can be determined by applying Ampere's right-handed screw rule. The induced current flows in counterclockwise direction to generate a negative magnetic flux (right to left) and opposes the increase of the positive magnetic flux (left to right).

● レンツの法則は，誘導電圧や誘導電流の向きを決定するための基本原理である．

図3・13(a)に示すように，磁石がコイルに近づくにつれて，コイルを鎖交する磁束が増加する．

ファラデーの電磁誘導の法則によれば，磁束の変化が生じると，起電力，

すなわち電流がコイルに誘導され，この電流がそれ自身の磁界をつくり出す．検流計は，電流の向きと大きさの両方を示している．

　レンツの法則によれば，電磁誘導によって発生する起電力は，コイルを鎖交する磁束の変化を妨げる向きに生じる．これは，コイルの左端がN極になり，磁石とコイルのN極が互いに反発することを意味する．

　コイル側の磁極がわかると，誘導電流の向きはアンペアの右ねじの法則を適用することによって決定することができる．誘導電流は，反時計方向に流れて負の磁束（右から左）を発生し，正の磁束（左から右）の増加を妨げる．

As the magnet moves away from the coil as shown in Fig.3・13(b), the magnetic flux interlinking across the coil decreases. The induced pole at the left end of the coil becomes a south pole according to Lenz's law.

The induced south pole will attract the north pole of the magnet to oppose the motion of the magnet moving away. For a south pole at the left end of the coil, then, the current flow will be reversed from the direction. Therefore, the induced current flows in clockwise direction to generate a positive magnetic flux and promotes the increase of the positive magnetic flux.

● 　図3・13(b)に示すように，磁石がコイルから遠ざかるにつれて，コイルを鎖交する磁束が減少する．コイルの左端の誘導された極は，レンツの法則によってS極となる．

　誘導されたS極は，遠ざかる磁石の動きに対抗して磁石のN極を引き付ける．コイルの左端がS極の場合，電流の流れは逆になる．したがって，誘導電流は，時計方向に流れて正の磁束を発生し，正の磁束の増加を促進する．

3.8　磁性体

磁界中に置かれた物質が磁化する現象を磁気誘導といい，その物

質を磁性体という．磁性体は，磁気構造の違いにより，いくつかに
分類される．

When a magnetic material like iron piece placed in the magnetic field of a permanent magnet, the iron piece is magnetized without actual contact with the permanent magnet. This phenomenon is known as magnetic induction.

In Fig.3･14, the magnetized iron piece has two poles. The end of the iron piece nearer the permanent magnet will be of opposite polarity and the farther end will be of the same polarity as that of the permanent magnet. Thus, since opposite poles attract, the iron piece will be attracted by the permanent magnet.

Induced magnetism is a temporary process. If the permanent magnet is removed, the iron piece will usually lose its induced magnetism.

● 　鉄片のような磁性体を，永久磁石の磁界中に置くと，鉄片は実際に永久
磁石と接触することなく磁化される．この現象は，磁気誘導として知られ
ている．
　　図3･14において，磁化された鉄片は，二つの極を有する．永久磁石に
近い方の鉄片の端部は，永久磁石と反対の極性になり，遠い方の端部は永
久磁石の極性と同じ極性になる．したがって，異極は引き付けあうので，
鉄片は永久磁石に引き付けられる．

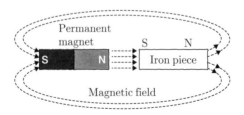

図3･14　Magnetic induction principles.

　誘導された磁気は，一時的なプロセスである．永久磁石が取り除かれると，鉄片は通常，その誘導された磁気を失う．

Magnetic materials are classified in five types of magnetic materials according to the magnetic behavior. They are ferromagnetic material, paramagnetic material, diamagnetic material, ferrimagnetic material, and antiferromagnetic material.

The materials which are strongly attracted by magnetic field are known as ferromagnetic materials. Ferromagnetic materials include iron, steel, nickel, and cobalt. They become strongly magnetized in the same direction as the magnetic field, with high values of permeability from 50 to 5 000. They retain their magnetic properties even after the external field has been removed.

The materials which are slightly attracted by magnettic field are known as paramagnetic materials. Paramagnetic materials include aluminum, platinum, manganese, and chromium. Their permeability is slightly more than 1. They lose their magnetic properties once the external magnetic field has been removed.

The materials which are weakly magnetized in the opposite direction to the applied magnetic field are known as diamagnetic materials. Diamagnetic materials include bismuth, antimony, copper, gold, silver, zinc, and mercury. Their permeability is less than 1. They are slightly magnetized when placed in a very strong magnetic field and act in the direction opposite to that of applied magnetic field.

Unlike ferromagnetic materials which is composed of metal, most of the ferrimagnetic materials are ceramic oxides. Ferrite and magnetic garnet are widely used as ferrimagnetic materials.

Finally, antiferromagnetic materials are known which are compounds of transition metals containing oxygen or sulfur. In antiferromagnetic materials, the magnetic moments consisting of vectors are aligned in opposite directions and equal in magnitude. Therefore, when an antiferromagnetic material is unmagnetized, its total magnetization is zero. In the presence of the strong magnetic field, antiferromagnetic materials are weakly magnetized in the direction of the magnetic field. Antiferromagnetic materials include MnO, FeO, CoO, NiO, Cr_2O_3, and MnS.

● 磁性体は，磁気的挙動に応じて5種類の磁性体に分類される．それらは，強磁性体，常磁性体，反磁性体，フェリ磁性体，および反強磁性体である．

　磁界によって強く引き付けられる物質は，強磁性体として知られている．強磁性体には，鉄，鋼，ニッケル，コバルトが含まれる．これらは磁界と同じ方向に強く磁化され，50～5 000の高い透磁率をもつ．それらは，外部磁界が取り除かれた後でも磁気特性を保持する．

　磁界によって弱く引き付けられる物質は，常磁性体として知られている．常磁性体には，アルミニウム，白金，マンガン，およびクロムが含まれる．これらの透磁率は，1よりわずかに大きい．それらは，外部磁界が取り除かれると磁気特性を失う．

　印加された磁界と逆の向きに弱く磁化される物質は，反磁性体として知られている．反磁性体には，ビスマス，アンチモン，銅，金，銀，亜鉛および水銀が含まれる．これらの透磁率は1未満である．それらは，非常に強い磁界中に置かれたときにわずかに磁化され，印加された磁界を打ち消すような向きに磁化される．

　金属からなる強磁性体とは異なり，フェリ磁性体のほとんどはセラミック酸化物である．フェライトと磁性ガーネットは，フェリ磁性体として広く用いられている．

　最後に，酸素または硫黄を含む遷移金属の化合物である反強磁性体が知られている．反強磁性体では，ベクトルからなる磁気モーメントは，反対方向に整列され，大きさは等しい．したがって，反強磁性体が磁化されて

いない場合，全体として磁化はゼロとなる．強い磁界が存在する場合，反
強磁性体は磁界の向きに弱く磁化される．反強磁性体には，酸化マンガン，
酸化第一鉄，酸化銅，酸化ニッケル，酸化クロムおよび硫化マンガンが含
まれる．

3.9　磁気回路

　環状の鉄心にコイルを巻き，それに電流を流すと鉄心内に磁束が
生じる．この磁束の通る閉路を磁気回路または磁路という．

A simple example of a magnetic circuit is shown in Fig.3·15.
When a current flows, magnetic flux is generated in the iron
core. The motive force that generates the magnetic flux is called
magnetomotive force. The magnetomotive force F is expressed by
the product NI of the number of turns and the current flowing in the
coil. The magnetomotive force F is often abbreviated to $m.m.f.$.

$$F = NI \text{ [A]}$$

where

F is the $m.m.f.$, in amperes [A]

I is the current, in amperes [A]

N is the number of turns (no units)

Ohm's law for magnetic circuit can be expressed as follows.

$$F = R_\mathrm{m}\phi$$

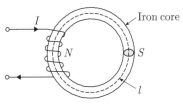

図 3・15　Magnetic circuit.

where R_m is the reluctance, and ϕ is the magnetic flux.

Hence, the reluctance R_m of a magnetic circuit can be expressed as follows.

$$R_m = \frac{F}{\phi} \text{ [A/Wb]}$$

$$R_m = \frac{l}{\mu S} = \frac{l}{\mu_0 \mu_r S} \text{ [A/Wb]}$$

where l is the average path length of the magnetic circuit

   S is the cross-sectional area of the core,

   μ_r is the relative permeability of the core.

Accordingly, the flux density B in the iron core is as follows.

$$B = \frac{\phi}{S} = \frac{\mu NI}{l} \text{ [T]} \tag{3-1}$$

The magnetic flux in a magnetic circuit is analogous to the electric current flowing in an electrical circuit. Similarly, magnetomotive force is similar to electromotive force.

Similarities between magnetic and electric circuits are shown in table.

Table

Magnetic circuit	Electric circuit
magnetomotive force F [A]	electromotive force E [V]
magnetic flux ϕ [Wb]	electric current I [A]
reluctance R_m [A/Wb]	resistance R [Ω]
permeability μ [H/m]	conductivity σ [S/m]

● 図3・15に磁気回路の簡単な例を示す．電流が流れると，鉄心に磁束が発生する．磁束を発生させる原動力は，起磁力と呼ばれる．起磁力 F は，

コイルの巻数と流れる電流の積 NI で表される。起磁力 F は，$m.m.f.$ と略されることが多い。

$$F = NI \text{ [A]}$$

ここで，

F は起磁力で，単位はアンペア [A]

I は電流で，単位はアンペア [A]

N は巻数（単位なし）

磁気回路のオームの法則は，次のように表すことができる。

$$F = R_\mathrm{m} \phi \text{ [A]}$$

ここで，R_m は磁気抵抗，ϕ は磁束である。

したがって，磁気回路の磁気抵抗 R_m は次のように表すことができる。

$$R_\mathrm{m} = \frac{F}{\phi} \text{ [A/Wb]}$$

$$R_\mathrm{m} = \frac{l}{\mu S} = \frac{l}{\mu_0 \mu_\mathrm{r} S} \text{ [A/Wb]}$$

ここで，l は磁気回路の平均磁路長，

S は鉄心の断面積，

μ_r は鉄心の比透磁率。

したがって，鉄心の磁束密度 B は次のようになる。

$$B = \frac{\phi}{S} = \frac{\mu NI}{l} \text{ [T]} \tag{3-1}$$

磁気回路における磁束は，電気回路を流れる電流と相似である。同様に，起磁力は，起電力に類似している。

磁気回路と電気回路との類似点を表に示す。

表

磁気回路	電気回路
起磁力 F [A]	起電力 E [V]
磁束 ϕ [Wb]	電流 I [A]
磁気抵抗 R_m [A/Wb]	電気抵抗 R [Ω]
透磁率 μ [H/m]	導電率 σ [S/m]

Example A toroidal core having a cross-sectional area of 1.2×10^{-4} m^2, and an average path length of 0.8 m has a coil of 8 500 turns wound uniformly around it. Calculate the current required to produce a flux of 3.20×10^{-6} Wb in the core. Assume the relative permeability of iron to be 5 000.

例題 断面積1.2×10^{-4} m^2, 平均磁路長0.8 mの環状鉄心は, その周りに均一に巻かれた巻数8 500のコイルを有している. 鉄心に3.20×10^{-6} Wbの磁束を発生させるのに必要な電流を計算しなさい. 鉄の比透磁率を5 000と仮定する.

Solution (答) Using Eq.(3-1),

(3-1)式を使って,

$$B = \frac{\phi}{S} = \frac{\mu NI}{l} \, [\text{T}]$$

$$\therefore \quad I = \frac{l\phi}{\mu NS} = \frac{l\phi}{\mu_0 \mu_\mathrm{r} NS}$$

$$= \frac{0.8 \times 3.20 \times 10^{-6}}{4\pi \times 10^{-7} \times 5\,000 \times 8\,500 \times 1.2 \times 10^{-4}}$$

$$= 3.99 \times 10^{-4} \text{ A}$$

3.10 *B-H* ヒステリシス曲線

鉄心などの強磁性体に巻いたコイルに電流を流すと, 磁束が生じる. 電流と磁束の代わりに, 磁束密度と磁界の強さとの関係を示す曲線を磁化曲線またはB-H曲線という.

ヒステリシスとは, 磁性体の状態が現在加えられている力だけでなく, それまでにたどってきた経過に依存する現象のことをいう. また, その磁性体が外部からの磁界の影響を受けて磁化されるとき, 磁束密度と磁界の変化の様子を表したものがB-Hヒステリシス曲線である.

Fig.3·16 shows a magnetic circuit for experimentally obtaining a *B-H* curve using a toroidal core having a coil with the number of turns N. When the current of I ampere is passed through the coil, magnetic flux is generated in the toroidal core.

Using the magnetic field strength H and the magnetic flux density B instead of the current and the magnetic flux, the magnetic flux density is a function of the magnetic field strength.

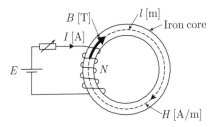

図3·16 Magnetic circuit.

● 図3·16は，巻数 N のコイルを有する環状鉄心を用いて，実験的に *B-H* 曲線を求める磁気回路を示す．電流 I アンペアがコイルを流れると，環状鉄心に磁束が生じる．
　電流と磁束の代わりに磁界の強さ H と磁束密度 B を用いると，磁束密度は磁界の強さの関数となる．

Fig.3·17 shows the relationship between magnetic field strength and the magnetic flux density when the magnetic field strength H is gradually increased from zero for the iron core. This is called as magnetization curve or the *B-H* curve.

As can be seen from Fig.3·17, *B-H* curve is composed of three regions: an initial nonlinear region, a linear region, and a saturation region. In the initial nonlinear region, B increases very slowly with respect to increase in H. In the linear region, B increases

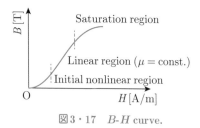

図 3・17 *B-H* curve.

in proportion to the increase of **H**. In the saturation region, **B** increases again very slowly with increase of **H**. Finally, **B** becomes almost constant with increasing **H**.

● 図3・17は，鉄心の磁界の強さ *H* をゼロから徐々に増加させたときの，磁界の強さと磁束密度との関係を示す．これは，磁化曲線または *B-H* 曲線と呼ばれる．

図3・17からわかるように，*B-H* 曲線は，初期非線形領域，線形領域，飽和領域の三つの領域から構成される．初期非線形領域では，*B* は *H* の増加に対して非常にゆっくりと増加する．線形領域では，*B* は *H* の増加に比例して増加する．飽和領域では，*B* は再び *H* の増加に対して非常にゆっくりと増加する．最後に，*B* は *H* の増加に対してほぼ一定になる．

Fig.3・18 is an experimental circuit for examining the relationship between the magnetic field strength and the magnitude and direction of a magnetic flux density by changing the magnitude and direction

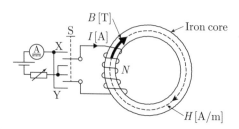

図 3・18 Experimental circuit to obtain hysteresis loop.

of a current flowing through a coil in a magnetic circuit of an iron core.

 First, when the switch S is placed in position X and the value of the current is gradually increased from zero, the magnetic field strength H and the magnetic flux density B gradually increase, and then the magnetic flux density rises to the maximum value called saturation. It is shown in Fig.3・19 as o-a.

 Next, when the value of the current is reduced to zero, both the magnetic field strength H and the magnetic flux density B decrease. It is shown as a-b. Here, when the switch S is moved from position X to position Y, the direction of the current changes, and the directions of the magnetic field and the magnetic flux density are opposite. As the value of the current is increased, the magnetic flux density B becomes zero. It is shown as b-c.

 Furthermore, as the value of the current is increased, both the magnetic field strength and the magnetic flux density decrease. It is shown as c-d. Next, as the value of the current is reduced to zero, the magnetic field strength H and the magnetic flux density B

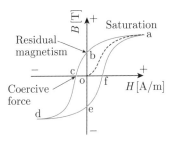

図 3・19 *B-H* hysteresis loop.

changes as shown by d-e.

Again, when the switch S is moved from position Y to position X, the direction of the current changes. As the value of the current is increased, both the magnetic field strength H and the magnetic flux density B increase. It is shown as e-f-a.

● 図3・18は，鉄心の磁気回路において，コイルに流れる電流の大きさと向きを変えて，磁界の強さと磁束密度の大きさと向きの関係を調べる実験回路である．

まず，スイッチSがX側にあり，電流の値がゼロから徐々に大きくなると，磁界の強さHと磁束密度Bは次第に増加し，磁束密度は飽和と呼ばれる最大値まで上昇する．これは，図3・19のo-aで示される．

次に，電流の値をゼロまで小さくすると，磁界の強さHと磁束密度Bはともに減少する．これは，a-bで示される．ここで，スイッチSがXからYに切り換えられると，電流の向きが変わり，したがって磁界と磁束密度の向きは反対になる．電流の値を大きくしていくと，磁束密度Bはゼロになる．これは，b-cで示される．

さらに，電流の値を大きくすると，磁界の強さと磁束密度はともに減少する．これは，c-dで示される．次に，電流の値をゼロまで小さくしていくと，磁界の強さHと磁束密度Bは，d-eで示されるように変化する．

再び，スイッチSがYからXに切り換えられると，電流の向きが変化する．その電流の値を大きくしていくと，磁界の強さHと磁束密度Bはともに増加する．これは，e-f-aで示される．

The closed path a-b-c-d-e-f-a shown in Fig.3・19 is called B-H hysteresis loop. When the magnetic field strength is reduced, the magnetic flux density also decreases. Actually, even when the magnetic field strength becomes zero, the magnetic flux density still has a positive value indicated by o-b in Fig.3・19. The value is called residual magnetism. Also, in order to cancel this residual magnetism and make the magnetic flux density in the iron core zero, a current in the opposite direction must flow. The magnetic field strength required to remove the residual magnetism from the iron core is called coercive force and is shown in Fig.3・19 as o-c.

As a magnetic material of the permanent magnet, a ferromagnetic material having high residual magnetism and coercive force is suitable. On the other hand, as a magnetic material used for an iron core of an electromagnet, a ferromagnetic material having a high magnetic flux density and a low coercive force at saturation is suitable.

● 図3・19に示した閉路a-b-c-d-e-f-aは*B-H*ヒステリシス曲線と呼ばれる．磁界の強さが低下すると，磁束密度も減少する．実際には，磁界の強さがゼロになっても，磁束密度は図3・19のo-bで示される正の値をなおも有している．この値は残留磁気と呼ばれる．さらに，この残留磁気を打ち消して鉄心中の磁束密度をゼロにするためには，逆向きの電流を流さなければならない．鉄心から残留磁気を除去するために必要な磁界の強さは保磁力と呼ばれ，図3・19のo-cで示される．

　永久磁石の磁性体としては，残留磁気と保磁力が大きい強磁性体が適している．一方，電磁石の鉄心に用いられる磁性体としては，飽和時の磁束密度が高く，保磁力が小さい強磁性体が適している．

3.11 熱電効果

　電気エネルギーと熱エネルギーの相互変換に作用する現象を熱電効果という．この効果には，ゼーベック効果，ペルチエ効果，トムソン効果などがある．

As shown in Fig.3・20, thermocouple is a pair of metals or semiconductors consisting of two dissimilar metals joined together at both ends. When one junction of the thermocouple is heated to generate a temperature difference between the two junctions, an electromotive force is generated and then a current flows. This

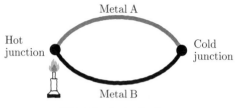

図3・20 Seebeck effect.

phenomenon is called the Seebeck effect, and the electromotive force generated is called thermoelectromotive force. In the thermocouple, the junction with the higher temperature is called the hot junction and the junction with lower temperature is called the cold junction.

● 図3・20に示すように，熱電対は，二つの異なる金属が両端で接合された一対の金属または半導体である．その熱電対の一方の接合点を加熱して二つの接合点の間に温度差を生じさせると，起電力が発生して電流が流れる．この現象をゼーベック効果といい，このとき発生する起電力を熱起電力という．熱電対で，温度の高い方の接合点を温接点，低い方の接合点を冷接点という．

In Fig.3.21, when two types of metals are joined and an electric current is passed, heat is absorbed at one junction and heat is released at the other junction. This phenomenon is called Peltier effect.

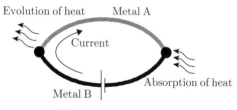

図3・21 Peltier effect.

When the direction of the current is reversed, the junction of heat absorption and the junction of release are exchanged.

● 図3・21において，2種類の金属を接合して電流を流すと，一方の接合点で熱の吸収，もう一方の接合点で熱の放出が起こる．この現象をペルチエ効果という．

電流の向きが反転すると，熱の吸収の接合点と放出の接合点が交換される．

In Fig.3.22, when a current is passed through a metal with a temperature difference, the phenomenon that generation or absorption of heat other than Joule's heat occurs is known as Thomson effect. Unlike the Seebeck effect and Peltier effect mentioned above, the Thomson effect is the thermoelectric effect in the single metal.

The quantity of heat Q generated or absorbed is proportional to the product of the current I and the temperature difference ΔT and expressed by the following equation.

$$Q = \theta \Delta T I \; [\mathrm{J}]$$

where, θ is a proportional coefficient, which is called Thomson coefficient.

● 図3・22において，温度差のある金属に電流を流すと，ジュール熱以外の熱の発生や吸収が起こる現象はトムソン効果として知られている．上記のゼーベック効果とペルチエ効果とは異なり，トムソン効果は同一金属で

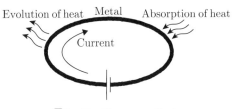

図3・22 Thomson effect.

の熱電効果である.

発生または吸収される熱量 Q は，電流 I と温度差 ΔT の積に比例し，以下の式で表される.

$$Q = \theta \Delta T I \text{ [J]}$$

ここで，θ は比例係数でトムソン係数と呼ばれる.

3.12 圧電効果

圧電効果とは，水晶やある種のセラミックに圧力または張力を加えると，それらの両端に電荷が発生する現象をいう．反対に，電極に電圧を印加することで変位を発生する現象を，逆圧電効果と呼ぶ.

電気的エネルギーと機械的エネルギーとの変換ができるので，発振器・スピーカ・点火装置・各種センサなどに応用される.

As shown in Fig.3·23, when pressure is applied to a quartz crystal or certain kinds of ceramics, the deviation of the position of the ions in it occurs. As a result, one end of the crystal has a positive electric charge and the other end has a negative electric charge. This is called electric polarization. That is, a phenomenon in which a voltage is generated according to strain by applying pressure to quartz or ceramic is called a piezoelectric effect.

On the other hand, when a voltage is applied to the quartz or the ceramic exhibiting a piezoelectric effect, physical strain is generated

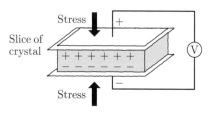

図3・23　Piezoelectric effect.

in the crystal. This is called inverse piezoelectric effect. The speaker for a mobile phone makes a use of this inverse piezoelectric effect to vibrate the air to produce a sound.

● 図3・23に示すように，水晶やある種のセラミックに圧力を加えると，結晶内のイオンの位置のずれが生じる．その結果，結晶の一端は正電荷を有し，他端は負電荷を有する．これは，電気分極と呼ばれる．すなわち，水晶やセラミックに圧力を加えて，ひずみに応じた電圧を発生させる現象を，圧電効果という．

一方，圧電効果を示す水晶やセラミックに電圧を印加すると，結晶に物理的なひずみが生じる．これを逆圧電効果という．携帯電話のスピーカは，この逆圧電効果を利用して，空気を振動させて音を出している．

A piezoelectric element is an electronic part that converts the pressure into electricity by utilizing piezoelectric effect or converts the voltage into vibration by using inverse piezoelectric effect. For example, a sensor using piezoelectric element converts a physical parameter such as acceleration or pressure into an electrical signal.

The structure of the piezoelectric element is comparatively simple, and fine motion and vibration are generated only by applying a voltage. Therefore, it can be miniaturized and is excellent as a part of a precision machine. An ultrasonic sensor is a device that generates ultrasonic waves using a piezoelectric element. When a potential difference is applied between the two terminals of the piezoelectric element, it expands and or contracts according to the direction of potential, and then ultrasonic waves are generated.

Products utilizing piezoelectric effect and ultrasonic waves are used in various fields such as daily life, medical field, and industrial field. For example, piezoelectric elements has been used for automatic ignition of electronic lighters and gas stoves. Furthermore, in the industrial field, it is used for positioning actuators and motors.

● 　圧電素子（ピエゾ素子）は，圧電効果を利用して圧力を電気に変換したり，逆圧電効果を利用して電圧を振動に変換する電子部品である．例えば，圧電素子を用いたセンサは，加速度や圧力などの物理的パラメータを電気信号に変換する．

　　圧電素子の構造は比較的単純であり，電圧を印加するだけで微細な動きや振動が発生する．したがって，小形化が可能であり，精密機械の部品として優れている．超音波センサは，圧電素子を利用して超音波を発生させるデバイスである．圧電素子の二つの端子間に電位差が印加されると，電位の方向に応じて伸縮し，超音波が発生する．

　　圧電効果や超音波を利用した製品は，日常生活，医療，産業などさまざまな分野で利用されている．例えば，圧電素子は，電子ライタやガスストーブの自動点火に使用されている．さらに，産業分野では，アクチュエータやモータの位置決めに利用されている．

4 交流回路

4.1 正弦波交流

交流とは，時間の経過とともに大きさと向きが周期的に変化する電気の流れである．ここでは正弦波交流でよく用いられる，サイクル，周期，周波数，ピーク値そしてピークトゥピーク値について述べる．

Alternating current is a current that continuously varies its magnitude and periodically reverses in direction.

In Fig.4・1, the waveform of a sine wave alternating voltage is represented by time on the horizontal axis, and voltage on the vertical axis. This is a change in polarity and voltage generated based on the principle of a generator.

● 交流とは，その大きさが連続的に変化し，周期的に向きが変わる電流である．

図4・1において，正弦波交流電圧の波形は，横軸に時間，縦軸に電圧で表されている．これは，発電機の原理に基づいて生成される極性および電

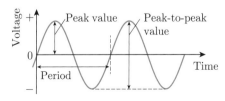

図4・1 Waveform of sine wave alternating voltage.

圧の変化である.

The waveform of sine wave alternating current is characterized by periodic repetition of the waveform.

A cycle is one complete waveform starting at a point and continuing until the same point on the pattern is reached. This is usually expressed from zero to zero or from peak to peak.

The time required to complete one cycle of the waveform is called the period of the wave, which is given the symbol T. The number of cycles per second is frequency and the symbol is f. In reference to frequency, the unit called Hertz (Hz) is used.

The frequency is the reciprocal of the period. The relation between the two is expressed as follows.

$$frequency = \frac{1}{period}$$

When the symbol f and the symbol T are used for frequency and for period respectively, these relations are expressed as follows.

$$f = \frac{1}{T}$$

The peak value or the maximum value indicates the highest value that the sine wave reaches. Moreover, peak-to-peak value of a sine wave is a value from the positive peak to the negative peak. For sine wave alternating voltage, the peak-to-peak value is represented by $V_{p\text{-}p}$.

● 正弦波交流波形は，波形が周期的に繰り返すのが特徴である.

　1サイクルは，波形上のある点から始まり，同じ点に到達するまで連続する全波形のことである．これは，通常，ゼロからゼロまたはピークから

ピークまでで表される．波形の1サイクルを完了するのに要する時間は，波形の周期と呼ばれ，記号 T で示される．

1秒当たりのサイクル数が周波数であり，記号は f である．周波数に関して，ヘルツ（Hz）と呼ばれる単位が用いられる．

周波数は周期の逆数である．両者の関係は以下のように表される．

$$\text{周波数} = \frac{1}{\text{周期}}$$

記号 f および記号 T を，それぞれ周波数および周期とすると，これらの関係は以下のように表される．

$$f = \frac{1}{T}$$

ピーク値または最大値は，正弦波が到達する最高値を示す．さらに，正弦波のピークトゥピーク値は，正のピークから負のピークまでの値である．正弦波交流電圧の場合，ピークトゥピーク値は $V_{\text{p-p}}$ で表される．

Example　What is the frequency of an AC (alternating current) waveform that has a period of 1 ms ?

例題　1 ms の周期をもつ交流波形の周波数はいくらか．

Solution (答)　$f = \dfrac{1}{T} = \dfrac{1}{1 \times 10^{-3}} = 1 \text{ kHz}$

4.2　交流の表し方

交流の大きさを表す方法として，瞬時値，平均値，実効値がある．

The sine wave alternating voltage is constantly varies in magnitude. The value of the amplitude at each time is called instantaneous value.

As mentioned above, period is the time taken to complete one full cycle. The distance travelled by the sine wave during this period is referred to as wavelength. For example, the velocity of electromagnetic wave in air or vacuum is 3×10^8 m/s , which is the

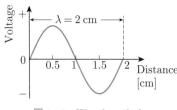

図 4・2　Wavelength λ.

speed of light. Therefore, the wavelength is expressed as follows.

$$\lambda = \frac{3 \times 10^8}{f}$$

where

λ = wavelength (meters)

f = frequency (Hz)

The fig.4・2 shows the wavelength for a radio wave with f of 15 GHz.

● 正弦波交流電圧は，絶えず大きさが変化する．各時刻の大きさの値は，瞬時値と呼ばれる．

上述したように，周期は，一つのサイクルを完了するのに要する時間である．この周期の間に，正弦波が進む距離は波長と呼ばれる．例えば，空気中または真空中の電磁波の速度は，光の速度である 3×10^8 m/s である．したがって，波長は以下のように表される．

$$\lambda = \frac{3 \times 10^8}{f}$$

ここで，

λは波長 (メートル)

f は周波数 (ヘルツ)

図4・2は，15 GHzの電波の波長を示している．

Example　Calculate the wavelength λ for a radio wave with frequency of 5 GHz.

例題 5 GHz の周波数をもつ電波の波長 λ を計算せよ.

Solution (答) $\lambda = \dfrac{3 \times 10^8}{5 \times 10^9} = 6 \text{ cm}$

The average value of the alternating voltage is defined for the positive half-cycle because the average of the voltage over one cycle is zero.

When V_m and V_{ave} are the peak and average values of the sine wave alternating voltage respectively, V_m and V_{ave} are related as follows.

$$V_m = \frac{\pi}{2} V_{ave}$$

● 交流電圧の平均値は,正の半サイクルに対して定義される.これは,1サイクルにわたる電圧の平均がゼロであるためである.
　V_m と V_{ave} がそれぞれ正弦波交流電圧のピーク値と平均値とすると,V_m と V_{ave} は次のように関係づけられる.

$$V_m = \frac{\pi}{2} V_{ave}$$

The most common method of stating the magnitude of an alternating voltage is by relating it to the DC voltage that will produce the same heating effect. This is called root-mean-square value or effective value, which is abbreviated to V_{rms}. V_m and V_{rms} are related as follows.

$$V_m = \sqrt{2}\, V_{rms}$$

For example, when a rms value is 100 V, the peak value is approximately 141 V.

● 交流電圧の大きさを示す最も一般的な方法は,同じ熱量を発生する直流電圧に関連づけることである.これは2乗平均平方根値または実効値と呼

ばれ，略して V_{rms} と書かれる．V_{m} と V_{rms} は次のように関係づけられる．
$$V_{\mathrm{m}} = \sqrt{2}\,V_{\mathrm{rms}}$$
例えば，実効値が 100 V の場合，ピーク値は約 141 V である．

4.3 正弦波交流の位相と位相差

周波数が同じ正弦波交流波形でも，その波形の時間方向の位置が波形によって異なる場合がある．正弦波交流波形の 1 サイクル内の時間的な位置を位相という．なお，位相に関して，位相角という用語が同義に用いられることがある．

The voltage waveform shown in Fig.4・3 is called a sine wave. The sine wave is can be expressed by the following equation.

$$v = V_{\mathrm{m}} \sin \omega t \,[\mathrm{V}]$$

where v is the instantaneous value of the sine wave at a given instant t.

V_{m} is the peak value of the sine wave.

ω is the angular frequency, expressed in radians per second. $\omega = 2\pi f$.

t is the time, expressed in seconds.

● 図 4・3 に示す電圧波形は，正弦波と呼ばれる．正弦波は，以下の式で表される．

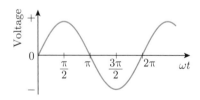

図 4・3　Sine wave with an initial phase of 0 degree.

$$v = V_\mathrm{m} \sin \omega t \, [\mathrm{V}]$$

ここで，vは与えられた時点tにおける正弦波の瞬時値である．

V_mは正弦波のピーク値である．

ωは角周波数で，ラジアン／秒で表される．$\omega = 2\pi f$.

tは時間で，秒で表される．

The general expression for a sine wave shown in Fig.4・4 will be given as follows.

$$v = V_\mathrm{m} \sin(\omega t + \theta) \, [\mathrm{V}]$$

where θ is the initial phase given in radians or degrees.

We can say that this sine wave is sin ωt that has been shifted toward the left by θ.

Since the sine wave is periodic wave with period 2π, the initial phase $\theta \pm 2\pi$ cannot be distinguished from θ. Therefore, we may restrict the range of θ so that it does not exceed 2π. In general, either of the following two ranges is chosen.

$$0 \leqq \theta < 2\pi, \text{ or } -\pi \leqq \theta < \pi.$$

● 図4・4に示す正弦波の一般的な式は，次のようになる．

$$v = V_\mathrm{m} \sin(\omega t + \theta) \, [\mathrm{V}]$$

ここで，θはラジアンまたは度で与えられる初期位相である．

この正弦波は，sin ωtをθだけ左にシフトしたものということができる．

正弦波は，周期2πの周期波であるため，初期位相$\theta \pm 2\pi$はθと区別できない．それゆえ，θの範囲が2πを超えないように制限される．一般に，

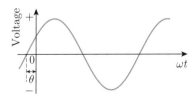

図4・4　Sine wave with an initial phase θ.

以下の二つの範囲のいずれかが選択される．
$$0 \leqq \theta < 2\pi, \quad \text{または} -\pi \leqq \theta < \pi.$$

Fig.4·5 shows a sine wave and a cosine wave. The two waves have the same frequency. Since sine wave A starts at zero, an initial phase is 0 radian. On the other hand, since cosine wave B starts at maximum, an initial phase is $\pi/2$ radians.

Phase difference is defined as the difference in initial phases between two waves having the same frequency. Therefore, the phase difference between sine wave A and cosine wave B is $\pi/2$ radians.

In addition, in order to clarify the temporal relationship between the two waves, it is often expressed by lead and delay in phase. Namely, cosine wave B leads the sine wave A by $\pi/2$ radians in phase.

A cosine function can be also written in terms of a sine function as

$$\cos \omega t = \sin\left(\omega t + \frac{\pi}{2}\right).$$

When the phase difference between two waves with the same frequency is 0, they are said to be in-phase.

● 図4·5は正弦波と余弦波を示している．二つの波は，同じ周波数を有する．正弦波Aは0から始まるので，初期位相は0 radである．一方，余弦

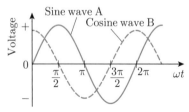

図4·5　Sine and cosine voltage waves.

波Bは最大で始まるので，初期位相はπ/2 radである．

位相差は，同じ周波数を有する二つの波の間の初期位相の差として定義される．したがって，正弦波Aと余弦波Bとの位相差はπ/2 radになる．

さらに，二つの波の時間的関係を明らかにするために，しばしば位相の進みと遅れで表現される．すなわち，余弦波Bは正弦波Aよりπ/2 rad進んでいる．

余弦関数は，$\cos \omega t = \sin\left(\omega t + \dfrac{\pi}{2}\right)$ のように正弦関数を用いて表すこともできる．

同じ周波数をもつ二つの波の位相差が0のとき，それらの波は同相であるという．

4.4 正弦波交流のフェーザ表示

正弦波交流の瞬時値は正弦関数で表されているが，グラフを用いて表すこともできる．フェーザ表示は，数学では極形式と呼ばれ，実効値で表した振幅と位相を複素ベクトル表示したものである．

In general, a sine wave is expressed by $v = V_m \sin(\omega t + \theta)$. However, in such a mathematical expression method, it may be difficult to visualize the phase difference between sine waves. In order to overcome this problem, we use a phasor diagram to graphically represent sine waves.

A phasor is a complex number that represents both magnitude and direction. The magnitude and argument are represented by effective value and phase difference, respectively. Phasor diagram is a graphical way of representing the magnitude and directional relationship between two sine waves with the same frequency.

● 一般に，正弦波は $v = V_m \sin(\omega t + \theta)$ で表される．しかし，このような表現法では，正弦波間の位相差を視覚化することが難しい場合がある．

この問題を解決するために，フェーザ図を用いて正弦波を図形的に表現する．

　フェーザは，大きさと方向の両方を表す複素数である．大きさと偏角は，それぞれ実効値と位相差で表される．フェーザ図は，同じ周波数をもつ二つの正弦波の間の大きさと方向性の関係をグラフで表現する方法である．

As an example, the two sine waves shown in Fig.4·6 are expressed by the following expressions.

$$v = V_m \sin \omega t \ [V]$$
$$i = I_m \sin(\omega t + \theta) \ [A]$$

The current i is leading the voltage v by phase difference of angle $\theta = \pi/6$. Fig.4·7 shows the phasor diagram.

As mentioned above, even if the two sine waves have different initial phases and different peak values, the phase difference can be expressed using a phasor diagram if they have the same frequency. In general, the phasor is represented by the effective value of the

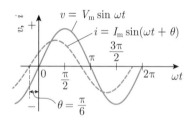

図 4・6　Phase difference between two sine waves.

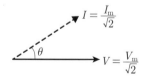

図 4・7　Phasor diagram.

sine wave. The reason is that the measured values by AC ammeter and AC voltmeter are mostly displayed as effective values. The effective values of voltage and current are represented by V and I respectively.

● 一例として，図4・6に示す二つの正弦波は，以下の式で表される．

$$v = V_\mathrm{m} \sin \omega t \ [\mathrm{V}]$$
$$i = I_\mathrm{m} \sin(\omega t + \theta) \ [\mathrm{A}]$$

電流 i は，電圧 v より位相差 $\theta = \pi/6$ だけ進んでいる．図4・7はそのフェーザ図を示す．

上述したように，二つの正弦波が異なる初期位相および異なるピーク値をもっていても，同じ周波数であればフェーザ図を用いて位相差を表すことができる．一般に，フェーザは，正弦波の実効値で表される．その理由は，交流電流計および交流電圧計による測定値の大部分が実効値で表示されるためである．電圧と電流の実効値は，それぞれ V と I で表される．

4.5 複素数表示

交流回路を解析するうえで，例えば電圧に対して振幅と位相の二つの成分がある．虚数を含んだ複素数を用いると，二つの成分を同時に扱えるので効率よく演算が可能になる．

When a complex number $\dot{z}$ can be expressed in the form $\dot{z} = x + \mathrm{j}y$, it is said to be in rectangular form, where x is the real part of $\dot{z}$, y is the imaginary part and both are real numbers. The imaginary unit j is defined by $\sqrt{-1}$. In mathematics, the symbol i is used to represent $\sqrt{-1}$, but in electricity, j is used to avoid confusion with i as the symbol for current.

As shown in Fig.4・8, a complex number in rectangular form can be represented as a point in two-dimensional plane with the real and imaginary parts corresponding to the real and imaginary axes

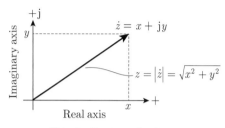

図 4・8　Rectangular form.

respectively.　This plane is called a complex plane or Gaussian plane.

As shown in this Fig., the magnitude of a complex number $\dot{z} = x + jy$ can be represented as follows.

$$z = \left| \dot{z} \right| = \sqrt{x^2 + y^2}$$

● 複素数 $\dot{z}$ が $\dot{z} = x + jy$ で表される場合，それは直交形式と呼ばれる．ここで，x は $\dot{z}$ の実部，y は虚部で，ともに実数である．虚数単位 j は $\sqrt{-1}$ で定義される．数学では，$\sqrt{-1}$ を表すのに記号 i が使われるが，電気では，電流の記号としての i との混乱を避けるため j が用いられる．

図 4・8 に示すように，直交形式の複素数は，実部と虚部がそれぞれ実軸と虚軸に対応する二次元平面の点として表すことができる．この平面は，複素平面またはガウス平面と呼ばれる．

この図に示すように，複素数 $\dot{z} = x + jy$ の大きさは次のように表すことができる．

$$z = \left| \dot{z} \right| = \sqrt{x^2 + y^2}$$

As shown in Fig.4·9, a complex number can also be represented in polar form, which consists of the distance from the origin and the angle measured counterclockwise from the positive real axis.

Representing a complex number $\dot{z}$ in polar form,

$$\dot{z} = x + jy = z(\cos \theta + j \sin \theta) = z \angle \theta$$

where

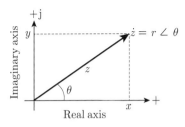

図 4・9　Polar form.

$$x = z \cos \theta$$
$$y = z \sin \theta$$
$$z = \sqrt{x^2 + y^2}$$
$$\theta = \tan^{-1} y/x$$

Polar form will help us understand complex numbers geometrically and are very useful for multiplication and division of complex numbers.

● 図 4・9 に示すように，複素数は，原点からの距離と正の実軸から反時計回りに測った角度からなる極形式で表すこともできる．

複素数 $\dot{z}$ を極形式で表現すると，

$$\dot{z} = x + \mathrm{j}y = z(\cos \theta + \mathrm{j} \sin \theta) = z \angle \theta$$

ここで

$$x = z \cos \theta$$
$$y = z \sin \theta$$
$$z = \sqrt{x^2 + y^2}$$
$$\theta = \tan^{-1} y/x$$

極形式は，複素数を幾何学的に理解するのに役立ち，複素数の乗算や除算に対して非常に有用である．

The complex conjugate of a complex number $\dot{z}$ denoted by $\overline{\dot{z}}$ is obtained by the replacement j → −j, so that $\overline{\dot{z}} = x - \mathrm{j}y$.

If we now calculate $\dot{z}\overline{\dot{z}}$, the following equation is obtained.

$$\dot{z}\overline{\dot{z}} = (x + \mathrm{j}y)(x - \mathrm{j}y) = x^2 + y^2$$

As a result, we see that $\dot{z}\overline{\dot{z}}$ is a positive real number and $\sqrt{\dot{z}\overline{\dot{z}}}$ is just the length of the vector $\dot{z}$ in the complex plane. Also, when the denominator of the fraction is a complex number, the denominator can be represented by a real number by using the property of this complex conjugate. This is called rationalization.

● $\overline{\dot{z}}$ で表される複素数 $\dot{z}$ の共役複素数は，j → −j の置き換えによって得られるので，$\overline{\dot{z}} = x - \mathrm{j}y$ となる．

ここで，$\dot{z}\overline{\dot{z}}$ を計算すると次の式が得られる．

$$\dot{z}\overline{\dot{z}} = (x + \mathrm{j}y)(x - \mathrm{j}y) = x^2 + y^2$$

　この結果，$\dot{z}\overline{\dot{z}}$ は正の実数，$\sqrt{\dot{z}\overline{\dot{z}}}$ は複素平面におけるベクトル $\dot{z}$ の長さであることがわかる．また，分数の分母が複素数である場合，この共役複素数の性質を利用すれば，分母を実数で表すことができる．これは有理化と呼ばれる．

4.6　交流抵抗回路

　ここでは，抵抗に正弦波交流が流れた場合，電流と抵抗の両端の電圧との位相関係を理解する．

　Fig.4・10(a) shows an alternating-current (AC) circuit consisting of an AC voltage source and resistance.

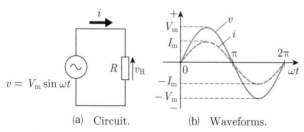

(a)　Circuit.　　　(b)　Waveforms.

図 4・10　An AC circuit with resistance R alone.

Let the alternating voltage applied across the circuit be given by the following equation.

$$v_R = V_m \sin \omega t \ [\text{V}] \tag{4-1}$$

Then the instantaneous value of current flowing through the resistance will be:

$$i = \frac{v_R}{R} = \frac{V_m \sin \omega t}{R} = I_m \sin \omega t \ [\text{A}] \tag{4-2}$$

From Eqs.(4-1) and (4-2), it is evident that the applied voltage and current are in phase in an AC circuit with resistance alone. This is shown in Fig.4·10(b).

Fig.4·11 shows the phasor diagram of an AC circuit with resistance alone. The phasors are represented by the effective value.

● 図4・10(a)は，交流電圧源と抵抗からなる交流回路を示す．
回路に印加される交流電圧は次の式で与えられる．

$$v_R = V_m \sin \omega t \ [\text{V}] \tag{4-1}$$

したがって，抵抗を流れる電流の瞬時値は次のようになる：

$$i = \frac{v_R}{R} = \frac{V_m \sin \omega t}{R} = I_m \sin \omega t \ [\text{A}] \tag{4-2}$$

(4-1)式および(4-2)式から，抵抗のみの交流回路において，印加電圧と電流は同相であることがわかる．これを第4.10(b)図に示す．

図4・11は，抵抗だけで構成された交流回路のフェーザ図を示している．フェーザは，実効値で表される．

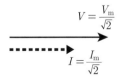

図4・11　Phasor diagram for the resistive circuit.

Example   As shown in Fig.4·12, when the source voltage v = 100 sin(ωt + π/6) [V] is connected to the resistance of 50 Ω, calculate the instantaneous current and effective value of current through the circuit.

例題   図4・12に示すように，電源電圧 v =100 sin(ωt + π/6) [V]が50 Ωの抵抗に接続されているとき，回路を流れる瞬時電流と電流の実効値を求めよ．

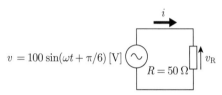

図4・12   A purely resistive AC circuit.

Solution（答）   The instantaneous voltage across the resistance is equal to the source voltage.

抵抗の両端の瞬時電圧は電源電圧に等しい．

$$\therefore \quad v_R = 100 \sin (\omega t + π/6) \text{ [V]}$$

Therefore, the instantaneous current flowing through the resistance is as follows.

したがって，抵抗に流れる瞬時電流は次のようになる．

$$i = \frac{v}{R} = \frac{100 \sin(\omega t + π/6)}{50}$$
$$= 2 \sin(\omega t + π/6) \, [\text{A}]$$

Also, the effective value of the current is as follows.

また，電流の実効値は以下のとおりである．

$$I = \frac{I_m}{\sqrt{2}} = \frac{2}{\sqrt{2}} = \sqrt{2} = 1.41 \text{ A}$$

4.7 交流コンデンサ回路

　交流電源とコンデンサだけで構成された回路は，交流コンデンサ回路や交流キャパシタンス回路と呼ばれる．この回路において，コンデンサの両端に生じる電圧と，コンデンサに流れる電流の関係を調べる.

　Fig.4·13(a) shows an AC circuit connected to capacitor only. When the alternating voltage given by equation (4-3) is applied to the circuit, charge of the capacitor at any instant of time is given in equation (4-4).

$$v = V_m \sin \omega t \text{ [V]} \tag{4-3}$$

$$q = Cv = CV_m \sin \omega t \tag{4-4}$$

where

　q = charge on the capacitor in coulombs

　C = capacitance of the capacitor in farads

　The current i flowing through the circuit is the flow of charge out of the capacitor per unit time and is expressed as follows.

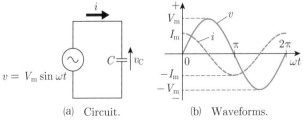

(a)　Circuit.　　　　(b)　Waveforms.

図 4·13　AC circuit with capacitance C.

$$i = \frac{dq}{dt} = \omega C V_m \cos \omega t = \omega C V_m \sin\left(\omega t + \frac{\pi}{2}\right)$$

$$= \frac{V_m}{\dfrac{1}{\omega C}} \sin\left(\omega t + \frac{\pi}{2}\right) = \frac{V_m}{X_C} \sin\left(\omega t + \frac{\pi}{2}\right) \text{[A]} \qquad (4\text{-}5)$$

where X_C is called capacitive reactance and the formula for it is as follows.

$$X_C = \frac{1}{\omega C} = \frac{1}{2\pi f C} \qquad (4\text{-}6)$$

When f is frequency in hertz and C is capacitance in farads, the unit of X_C is ohm. The capacitive reactance decreases with increasing frequency.

In equation (4-5), when $\sin(\omega t + \pi/2)$ is equal to one, the value of current will be at maximum.

Therefore,

$$I_m = \frac{V_m}{X_C} \qquad (4\text{-}7)$$

From equations (4-5) and (4-7), we will get the following equation.

$$i = I_m \sin\left(\omega t + \frac{\pi}{2}\right) \qquad (4\text{-}8)$$

From equation (4-3) of the source voltage and equation (4-8) of the current flowing through the capacitor, it is clear that the current leads the voltage by $\pi/2$. This is shown in Fig.4·13(b).

Both current and voltage are expressed as effective values as follows.

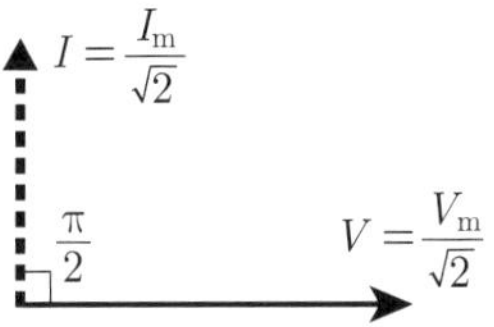

$$I = \frac{I_\mathrm{m}}{\sqrt{2}}$$

$$\frac{\pi}{2}$$

$$V = \frac{V_\mathrm{m}}{\sqrt{2}}$$

図 4・14 Phasor diagram for an AC circuit with capacitance only.

$$I = \frac{V}{X_\mathrm{C}} \tag{4-9}$$

Fig. 4・14 shows a phasor diagram for an AC circuit consisting only of a capacitor. The phasors are represented by the effective value.

● 図 4・13(a)は, コンデンサだけが接続された交流回路を示している. (4-3)式で与えられる交流電圧をこの回路に印加すると, 任意の時刻でのコンデンサの電荷は (4-4)式で与えられる.

$$v = V_\mathrm{m} \sin \omega t \ [\mathrm{V}] \tag{4-3}$$

$$q = Cv = CV_\mathrm{m} \sin \omega t \tag{4-4}$$

ここで, q はコンデンサの電荷で単位はクーロン, C はコンデンサの静電容量で単位はファラドである.

回路を流れる電流 i は, 単位時間当たりのコンデンサからの電荷の流れであり, 以下のように表される.

$$i = \frac{\mathrm{d}q}{\mathrm{d}t} = \omega C V_\mathrm{m} \cos \omega t = \omega C V_\mathrm{m} \sin\left(\omega t + \frac{\pi}{2}\right)$$

$$= \frac{V_\mathrm{m}}{\dfrac{1}{\omega C}} \sin\left(\omega t + \frac{\pi}{2}\right) = \frac{V_\mathrm{m}}{X_\mathrm{C}} \sin\left(\omega t + \frac{\pi}{2}\right) [\mathrm{A}] \tag{4-5}$$

ここで, X_C は容量性リアクタンスと呼ばれ, その式は次のようになる.

$$X_\mathrm{C} = \frac{1}{\omega C} = \frac{1}{2\pi f C} \tag{4-6}$$

f がヘルツ単位の周波数, C がファラド単位の静電容量の場合, X_C の単位はオームである. 容量性リアクタンスは, 周波数の増加とともに減少する.

(4-5)式において, $\sin(\omega t + \pi/2)$ が 1 に等しいとき, 電流の値は最大になる.

したがって,

$$I_\mathrm{m} = \frac{V_\mathrm{m}}{X_\mathrm{C}} \tag{4-7}$$

(4-5)式および(4-7)式から，以下の式が得られる．

$$i = I_\mathrm{m} \sin\left(\omega t + \frac{\pi}{2}\right) \tag{4-8}$$

電源電圧の(4-3)式とコンデンサを流れる電流の(4-8)式から，電流は電圧より$\pi/2$だけ進んでいることが明らかである．これを図4・13(b)に示す．電流，電圧とも実効値で表すと以下のようになる．

$$I = \frac{V}{X_\mathrm{C}} \tag{4-9}$$

図4・14は，コンデンサのみからなる交流回路のフェーザ図を示している．フェーザは，実効値で表されている．

Example　In Fig.4·13(a), when the effective value of the applied source voltage is 1.50×10^2 V and C is 8 μF, calculate the capacitive reactance and the effective value of the current flowing through the circuit whose frequency is 50 Hz.

例題　図4・13(a)において，印加された電源電圧の実効値が1.50×10^2 V，Cが8 μFのとき，周波数50 Hzの回路における容量性リアクタンスと回路を流れる電流の実効値を求めよ．

Solution(答)　Substitute $C = 8$ μF and $f = 50$ Hz into Eq.(4-6) to get the capacitive reactance.

容量性リアクタンスを求めるために，$C = 8$ μF と $f = 50$ Hz を(4-6)式に代入する．

$$X_\mathrm{C} = \frac{1}{2\pi f C} = \frac{1}{2\pi \times 50 \times 8 \times 10^{-6}} \fallingdotseq 398 \ \Omega$$

Substitute $V = 1.50 \times 10^2$ V and $X_\mathrm{C} = 398$ Ω into Eq.(4-9) to get the current.

電流を求めるために，$V = 1.50 \times 10^2$ V と $X_C = 398$ Ω を (4-9) 式に代入する．

$$I = \frac{V}{X_C} = \frac{1.50 \times 10^2}{398} \fallingdotseq 0.38 \text{ A}$$

4.8　交流コイル回路

交流電源とコイルだけで構成された回路は，交流コイル回路や交流インダクタンス回路と呼ばれる．この回路において，コイルの両端に生じる電圧と，コイルに流れる電流の関係を調べる．

Fig.4·15(a) shows an AC circuit connected to inductor only. The alternating voltage applied to the circuit is given by Eq.(4-10).

$$v = V_\text{m} \sin \omega t \text{ [V]} \tag{4-10}$$

As a result, current i flows in the circuit and then the induced electromotive force represented by the Eq.(4-11) is generated across the inductor.

$$v_\text{L} = -L \frac{\mathrm{d}i}{\mathrm{d}t} \text{[V]} \tag{4-11}$$

Note that the negative sign indicates that the induced electromotive force v_L opposes the change in current through the inductor per unit

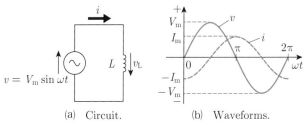

(a)　Circuit.　　　　(b)　Waveforms.

図 4·15　AC circuit with inductance L.

time. The induced electromotive force is equal and opposite to the applied voltage v. Therefore, the relationship between v and v_L is as follows.

$$v = -v_L \tag{4-12}$$

From Eqs.(4-11) and (4-12), we will get the following equation.

$$v = L\frac{di}{dt}[V] \tag{4-13}$$

$$\therefore \ V_m \sin \omega t = L\frac{di}{dt} \tag{4-14}$$

Rewrite Eq.(4-14) as follows.

$$di = \frac{V_m}{L} \sin \omega t\, dt \tag{4-15}$$

Integrating both sides of Eq.(4-15), we will get

$$\int di = \int \frac{V_m}{L} \sin \omega t\, dt$$

$$i = \frac{V_m}{\omega L}(-\cos \omega t) = \frac{V_m}{\omega L} \sin\left(\omega t - \frac{\pi}{2}\right)$$

$$= \frac{V_m}{X_L} \sin\left(\omega t - \frac{\pi}{2}\right)[A] \tag{4-16}$$

where X_L is called inductive reactance and the formula for it is as follows.

$$X_L = \omega L = 2\pi f L \tag{4-17}$$

When f is frequency in hertz and L is inductance in henrys, the unit of X_L is ohm. The inductive reactance increases with an increase in frequency.

The value of current will be maximum when $\sin(\omega t - \pi/2) = 1$.

Therefore,

$$I_m = \frac{V_m}{X_L} \tag{4-18}$$

From Eqs.(4-16) and (4-18), we will get the following equation.

$$i = I_m \sin\left(\omega t - \frac{\pi}{2}\right) \tag{4-19}$$

From Eq.(4-10) of the source voltage and Eq.(4-19) of the current flowing through the inductor, it is clear that the current lags the voltage by $\pi/2$. This is shown in Fig.4·15(b).

Both current and voltage are expressed as effective values as follows.

$$I = \frac{V}{X_L} \tag{4-20}$$

Fig.4·16 shows a phasor diagram for an AC circuit consisting only of an inductor. The phasors are represented by the effective value.

● 図4·15(a)は，コイルだけが接続された交流回路を示している．回路に加えられる交流電圧は，(4-10)式で与えられる．

$$v = V_m \sin \omega t \text{ [V]} \tag{4-10}$$

その結果，回路に交流電流 i が流れ，(4-11)式で表される誘導起電力がコイルの両端に発生する．

$$v_L = -L\frac{di}{dt} \text{ [V]} \tag{4-11}$$

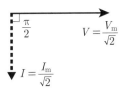

図4·16 Phasor diagram for an AC circuit with inductance only.

なお，留意すべきは，負の符号は，コイルを流れる単位時間当たりの電流の変化を妨げる向きに，誘導起電力 v_L が発生することを示していることである．誘導起電力は，印加電圧と等しく，向きは逆である．したがって，v と v_L の関係は次式のとおりである．

$$v = -v_L \tag{4-12}$$

(4-11)式および(4-12)式から，以下の式が得られる．

$$v = L \frac{\mathrm{d}i}{\mathrm{d}t} \, [\mathrm{V}] \tag{4-13}$$

$$\therefore \quad V_m \sin \omega t = L \frac{\mathrm{d}i}{\mathrm{d}t} \tag{4-14}$$

(4-14)式を次のように書き直す．

$$\mathrm{d}i = \frac{V_m}{L} \sin \omega t \, \mathrm{d}t \tag{4-15}$$

(4-15)式の両辺を積分することにより

$$\int \mathrm{d}i = \int \frac{V_m}{L} \sin \omega t \, \mathrm{d}t$$

$$i = \frac{V_m}{\omega L}(-\cos \omega t) = \frac{V_m}{\omega L} \sin\left(\omega t - \frac{\pi}{2}\right)$$

$$= \frac{V_m}{X_L} \sin\left(\omega t - \frac{\pi}{2}\right) [\mathrm{A}] \tag{4-16}$$

が得られる．

ここで，X_L は誘導性リアクタンスと呼ばれ，その式は次のようになる．

$$X_L = \omega L = 2\pi f L \tag{4-17}$$

f がヘルツを単位とする周波数，L がヘンリーを単位とするインダクタンスとすると，X_L の単位はオームである．誘導性リアクタンスは，周波数の増加とともに増加する．

電流の値は，$\sin(\omega t - \pi/2) = 1$ のとき最大である．したがって，

$$I_m = \frac{V_m}{X_L} \, [\mathrm{A}] \tag{4-18}$$

となる．

(4-16)式および(4-18)式から，以下の式が得られる．

$$i = I_m \sin\left(\omega t - \frac{\pi}{2}\right) \tag{4-19}$$

電源電圧の式(4-10)とコイルを流れる電流の(4-19)式から，電流は電圧よりπ/2だけ遅れていることが明らかである．これを図4・15(b)に示す．電流，電圧とも実効値で表すと，以下のようになる．

$$I = \frac{V}{X_{\mathrm{L}}} \, [\mathrm{A}] \tag{4-20}$$

図4・16は，コイルのみからなる交流回路のフェーザ図を示している．フェーザは，実効値で表されている．

Example In Fig.4·15(a), when the effective value of the applied source voltage is 1.50×10^2 V and L is 25 mH, calculate the inductive reactance and the effective value of the current in the circuit whose frequency is 50 Hz.

例題 図4・15(a)において，印加された電源電圧の実効値が1.50×10^2 V，Lが25 mHのとき，周波数50 Hzにおける誘導性リアクタンスと回路を流れる電流の実効値を求めよ．

Solution（答） Substitute $L = 25 \times 10^{-3}$ H and $f = 50$ Hz into equation(4-17) to get the inductive reactance.

誘導性リアクタンスを求めるために，$L = 25 \times 10^{-3}$ Hと$f = 50$ Hzを(4-17)式に代入する．

$$X_{\mathrm{L}} = \omega L = 2\pi f L \fallingdotseq 2 \times 3.14 \times 50 \times 25 \times 10^{-3}$$
$$= 7.85 \ \Omega$$

Substitute $E = 1.50 \times 10^2$ V and $X_{\mathrm{L}} = 7.85 \ \Omega$ into equation(4-20) to get the current.

電流を求めるために，$V = 1.50 \times 10^2$ Vと$X_{\mathrm{L}} = 7.85 \ \Omega$を(4-20)式に代入する．

$$I = \frac{V}{X_{\mathrm{L}}} = \frac{1.50 \times 10^2}{7.85} \fallingdotseq 19.1 \ \mathrm{A}$$

4.9 *RC* 直列回路

抵抗とコンデンサを直列に接続して交流電源につないだ回路は，*RC* 直列回路と呼ばれる.

Fig.4.17(a) shows a circuit consisting of resistance in series with capacitor, connected across an AC voltage source at a frequency of f [Hz]. This circuit is called resistance-capacitor series circuit. In this Fig., $\dot{V}_\mathrm{R}$ is the voltage across the resistance and $\dot{V}_\mathrm{C}$ is the voltage across the capacitor. The sum of $\dot{V}_\mathrm{R}$ and $\dot{V}_\mathrm{C}$ is equal to the source voltage $\dot{V}$ applied to the circuit. $\dot{I}$ is the current flowing through the circuit. All of these voltages and current are vector quantities.

By the application of Kirchhoff's voltage law to this circuit, we get

$$\dot{V} = \dot{V}_\mathrm{R} + \dot{V}_\mathrm{C}$$
$$\dot{V}_\mathrm{R} = \dot{I} R$$
$$\dot{V}_\mathrm{C} = \dot{I} X_\mathrm{C}$$

where X_C is capacitive reactance in ohms.

● 図 4・17(a)は，周波数 f [Hz] の交流電圧源に接続された，抵抗とコンデンサの直列接続からなる回路を示す．この回路は，*RC* 直列回路と呼ばれる．この図で，$\dot{V}_\mathrm{R}$ は抵抗両端の電圧，$\dot{V}_\mathrm{C}$ はコンデンサ両端の電圧であ

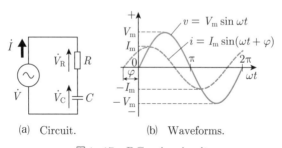

(a)　Circuit.　　　　(b)　Waveforms.

図 4・17　*RC* series circuit.

る．$\dot{V}_R$ と $\dot{V}_C$ の合計は，回路に印加される電源電圧 $\dot{V}$ に等しい．$\dot{I}$ は回路を流れる電流である．これらの電圧と電流は，すべてベクトル量である．

　この回路にキルヒホッフの電圧則を適用することにより，

$$\dot{V} = \dot{V}_R + \dot{V}_C$$
$$\dot{V}_R = \dot{I}_R$$
$$\dot{V}_C = \dot{I}X_C$$

ここで，X_C はオームを単位とする容量性リアクタンスである．

Fig.4·18 shows a phasor diagram for an RC series circuit. The phasor OA represents current $\dot{I}$. The phasor OB is the voltage $\dot{V}_R$ across the resistance which is in phase with the current. The phasor OC shows the voltage $\dot{V}_C$ across the capacitor. $\dot{V}_C$ lags $\dot{I}$ by $\pi/2$ rad. The phasor OD is the phasor sum of $\dot{V}_R$ and $\dot{V}_C$. The phasor OD represents the total of the terminal voltages of the circuit and is equal to the applied voltage $\dot{V}$. Therefore, in an RC series circuit, current $\dot{I}$ leads the total terminal voltage $\dot{V}$ by phase difference φ. These phasors are represented by the effective values.

As $\dot{V}_R = \dot{I}_R$ and $\dot{V}_C = \dot{I}X_C = \dot{I}/2\pi fC$, the total terminal voltage can be represented as follows.

$$V = \sqrt{V_R{}^2 + V_C{}^2} = \sqrt{(IR)^2 + (IX_C)^2} = I\sqrt{R^2 + X_C{}^2}$$

Therefore, the current flowing through the circuit,

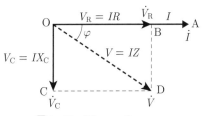

図 4・18　Phasor diagram.

127

$$I = \frac{V}{\sqrt{R^2 + X_C{}^2}} = \frac{V}{Z}$$

where

$$Z = \sqrt{R^2 + X_C{}^2}$$

Z is the impedance in RC series circuit, which is measured in ohms.

Current flowing through the circuit leads the total terminal voltage by an angle φ, which is given by Fig.4・18.

$$\varphi = -\tan^{-1}\left(\frac{V_C}{V_R}\right) = -\tan^{-1}\left(\frac{IX_C}{IR}\right) = -\tan^{-1}\left(\frac{X_C}{R}\right)$$

where the negative sign indicates that the voltage lags behind the current. The phase relationship between the instantaneous value v of the applied voltage and the instantaneous value i of the current is shown in the waveform in Fig.4・17(b).

● 図4・18は，RC直列回路のフェーザ図を表している．フェーザOAは，電流$\dot{I}$を表している．フェーザOBは，電流と同相の抵抗両端の電圧$\dot{V}_R$である．フェーザOCは，コンデンサ両端の電圧$\dot{V}_C$を表している．$\dot{V}_C$は，$\dot{I}$より$\pi/2$ rad遅れている．フェーザODは，$\dot{V}_R$と$\dot{V}_C$のフェーザの和である．フェーザODは，回路の端子電圧の合計を表し，印加電圧$\dot{V}$に等しい．それゆえ，RC直列回路では，電流$\dot{I}$は全体の端子電圧$\dot{V}$より角度φだけ位相が進んでいる．これらのフェーザは，実効値で表されている．

$\dot{V}_R = \dot{I}R$ および $\dot{V}_C = \dot{I}X_C = \dot{I}/2\pi fC$ であるので，全体の端子電圧は以下のように表すことができる．

$$V = \sqrt{V_R{}^2 + V_C{}^2} = \sqrt{(IR)^2 + (IX_C)^2} = I\sqrt{R^2 + X_C{}^2}$$

それゆえ，回路を流れる電流は

$$I = \frac{V}{\sqrt{R^2 + X_C{}^2}} = \frac{V}{Z}$$

となる．ここで，

$$Z = \sqrt{R^2 + X_C{}^2}$$

である．Zは，RC直列回路におけるインピーダンスで，単位はオームで

ある．

回路を流れる電流は，全体の端子電圧より角度 φ だけ進み，図4・18で与えられる．

$$\varphi = -\tan^{-1}\left(\frac{V_C}{V_R}\right) = -\tan^{-1}\left(\frac{IX_C}{IR}\right) = -\tan^{-1}\left(\frac{X_C}{R}\right)$$

ここで，負号は，電圧が電流より遅れていることを示す．印加電圧の瞬時値 v と電流の瞬時値 i の位相関係を図4.17(b)の波形に示す．

4.10 *RLC* 直列回路

抵抗，コイル，コンデンサを直列に接続して交流電源につないだ回路は，RLC 直列回路と呼ばれる．

Fig.4.19 shows a circuit consisting of a resistance, capacitor, and inductor in series across an AC voltage source. This circuit is called RLC series circuit. In this Fig., $\dot{V}_R$ is the voltage across the resistance, $\dot{V}_L$ is the voltage across the inductor and $\dot{V}_C$ is the voltage across the capacitor. The sum of $\dot{V}_R$, $\dot{V}_L$, and $\dot{V}_C$ is equal to the source voltage $\dot{V}$ applied to the circuit. The same current $\dot{I}$ flows each circuit element. All of these voltages and current are vector quantities.

Applying Kirchhoff's voltage law at the circuit, we will get the following equation.

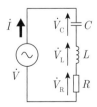

図4・19 *RLC* series circuit.

$$\dot{V} = \dot{V}_R + \dot{V}_L + \dot{V}_C$$
$$\dot{V}_R = \dot{I}R$$
$$\dot{V}_L = \dot{I}X_L$$
$$\dot{V}_C = \dot{I}X_C$$

where X_L is inductive reactance, and X_C is capacitive reactance, both in ohms.

$\dot{V}_R$ is in phase with $\dot{I}$. On the other hand, $\dot{V}_L$ leads $\dot{I}$ by $\pi/2$ rad, and $\dot{V}_C$ lags $\dot{I}$ by $\pi/2$ rad. Therefore, $\dot{V}_L$ and $\dot{V}_C$ are opposite to each other.

RLC series circuit acts as an RC series circuit if $X_C > X_L$, and acts as an RL series circuit if $X_C < X_L$.

● 図4・19は，交流電圧源に直列に抵抗，コンデンサ，およびコイルからなる回路を示している．この回路は，RLC直列回路と呼ばれる．この図で，$\dot{V}_R$は抵抗両端の電圧，$\dot{V}_L$はコイル両端の電圧，そして$\dot{V}_C$はコンデンサ両端の電圧である．$\dot{V}_R$，$\dot{V}_L$，$\dot{V}_C$の合計は，回路に印加される電源電圧$\dot{V}$に等しい．同じ電流$\dot{I}$が各回路素子を流れる．これらの電圧と電流は，すべてベクトル量である．

この回路にキルヒホッフの電圧則を適用すると，次式が得られる．
$$\dot{V} = \dot{V}_R + \dot{V}_L + \dot{V}_C$$
$$\dot{V}_R = \dot{I}R$$
$$\dot{V}_L = \dot{I}X_L$$
$$\dot{V}_C = \dot{I}X_C$$
ここで，X_Lは誘導性リアクタンス，X_Cは容量性リアクタンスで，両方ともオームである．

$\dot{V}_R$は$\dot{I}$と同相である．一方，$\dot{V}_L$は$\dot{I}$より$\pi/2$ rad 進み，$\dot{V}_C$は$\dot{I}$より$\pi/2$ rad 遅れる．それゆえ，$\dot{V}_L$と$\dot{V}_C$は互いに逆向きとなる．

RLC直列回路は，$X_C > X_L$の場合はRC直列回路として機能し，$X_C < X_L$の場合はRL直列回路の働きをする．

As an example, a phasor diagram in the case of $X_C > X_L$ is shown in Fig.4・20. From this phasor diagram, the total terminal voltage V

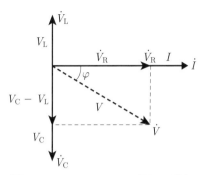

図 4・20 Phasor diagram($X_C > X_L$).

is expressed as follows.

$$V = \sqrt{V_R{}^2 + (V_C - V_L)^2} = \sqrt{(IR)^2 + (IX_C - IX_L)^2}$$
$$= I\sqrt{R^2 + (X_C - X_L)^2}$$
$$\therefore \ I = \frac{V}{\sqrt{R^2 + (X_C - X_L)^2}} = \frac{V}{Z}$$

where

$$Z = \sqrt{R^2 + (X_C - X_L)^2}$$

Z is known as combined impedance of the RLC series circuit and acts to oppose the flow of current.

From this phasor diagram, the value of phase difference will be

$$\varphi = -\tan^{-1}\left(\frac{V_C - V_L}{V_R}\right) = -\tan^{-1}\left(\frac{X_C - X_L}{R}\right)$$

● 例として，$X_C > X_L$ の場合のフェーザ図を図 4・20 に示す．このフェーザ図から，全体の端子電圧 V は次のように表される．

$$V = \sqrt{V_R{}^2 + (V_C - V_L)^2} = \sqrt{(IR)^2 + (IX_C - IX_L)^2}$$
$$= I\sqrt{R^2 + (X_C - X_L)^2}$$
$$\therefore \ I = \frac{V}{\sqrt{R^2 + (X_C - X_L)^2}} = \frac{V}{Z}$$

ここで，
$$Z = \sqrt{R^2 + (X_C - X_L)^2}$$

Z は，RLC 直列回路の合成インピーダンスとして知られ，電流の流れを妨げるように作用する.

このフェーザ図から，位相差の値は

$$\varphi = -\tan^{-1}\left(\frac{V_C - V_L}{V_R}\right) = -\tan^{-1}\left(\frac{X_C - X_L}{R}\right)$$

となる.

4.11 記号法

交流回路の計算で，複素数を用いて表す方法を記号法（symbolic method）という．記号法を用いると，電圧や電流などの計算が加算や乗算で簡単に表現できる．ここでは，RC 直列回路と RL 直列回路について解析を行うことにする．ここで，例えば $\dot{V}$ は複素成分の含まれる V であることを意味する.

(i) *RC* 直列回路（*RC* series circuit）

When a current flowing in the RC series circuit of Fig.4·21 is represented by a complex number $\dot{I}$ [A], the voltage $\dot{V}_R$ [V] across the resistor and the voltage $\dot{V}_C$ [V] across the capacitor are expressed as follows.

$$\dot{V}_R = \dot{I}R$$

$$\dot{V}_C = \frac{1}{j\omega C}\dot{I} = \frac{j}{j^2\omega C}\dot{I} = -j\frac{1}{\omega C}\dot{I}$$

where j is $\sqrt{-1}$

The source voltage $\dot{V}$ is expressed as follows using $\dot{V}_R$ and $\dot{V}_C$.

$$\dot{V} = \dot{V}_R + \dot{V}_C = \dot{I}R - j\frac{1}{\omega C}\dot{I} = \left(R - j\frac{1}{\omega C}\right)\dot{I}$$

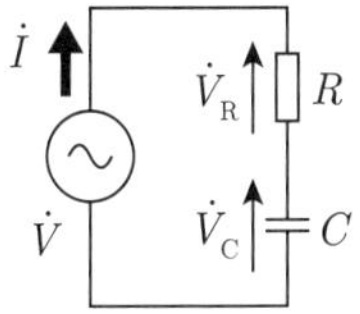

図 4・21　*RC* series circuit.

Therefore, the combined impedance $\dot{Z}$ [Ω] of the RC series circuit is expressed as follows.

$$\dot{Z} = \frac{\dot{V}}{\dot{I}} = R - j\frac{1}{\omega C} = R - jX_C$$

● 図 4・21 の RC 直列回路に流れる電流を複素数 $\dot{I}$ [A] で表すと，抵抗両端の電圧 $\dot{V}_R$ [V] とコンデンサ両端の電圧 $\dot{V}_C$ [V] は以下のように表される．

$$\dot{V}_R = \dot{I}R$$

$$\dot{V}_C = \frac{1}{j\omega C}\dot{I} = \frac{j}{j^2\omega C}\dot{I} = -j\frac{1}{\omega C}\dot{I}$$

ここで，j は $\sqrt{-1}$ である．

全体の端子電圧 $\dot{V}$ は，$\dot{V}_R$ と $\dot{V}_C$ を用いて次のように表される．

$$\dot{V} = \dot{V}_R + \dot{V}_C = \dot{I}R - j\frac{1}{\omega C}\dot{I} = \left(R - j\frac{1}{\omega C}\right)\dot{I}$$

したがって，RC 直列回路の合成インピーダンス $\dot{Z}$ [Ω] は以下のように表される．

$$\dot{Z} = \frac{\dot{V}}{\dot{I}} = R - j\frac{1}{\omega C} = R - jX_C$$

⒤ *RL* 直列回路（*RL* series circuit）

When the current flowing through the RL series circuit in Fig.4·22 is represented by a complex number $\dot{I}$ [A], the voltage $\dot{V}_R$ [V] across the resistance and the voltage $\dot{V}_L$ [V] across the inductor are expressed as follows.

$$\dot{V}_R = \dot{I}R$$

$$\dot{V}_L = j\omega L\dot{I}$$

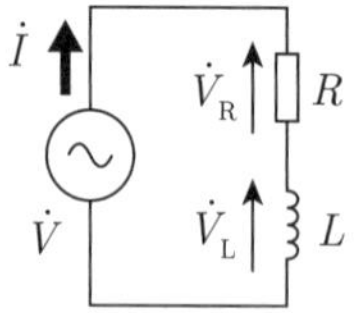

図 4・22　*RL* series circuit.

The total terminal voltage is

$$\dot{V} = \dot{V}_\mathrm{R} + \dot{V}_\mathrm{L} = \dot{I}R + \mathrm{j}\omega L\dot{I} = (R + \mathrm{j}\omega L)\dot{I}$$

Thus the combined impedance $\dot{Z}$ [Ω] of the *RL* series circuit is expressed as follows:

$$\dot{Z} = \frac{\dot{V}}{\dot{I}} = R + \mathrm{j}\omega L = R + \mathrm{j}X_\mathrm{L}$$

● 図 4・22 の *RL* 直列回路に流れる電流を複素数 $\dot{I}$ [A] で表すと，抵抗両端の電圧 $\dot{V}_\mathrm{R}$ [V] とコイル両端の電圧 $\dot{V}_\mathrm{L}$ [V] は以下のように表される.

$$\dot{V}_\mathrm{R} = \dot{I}R$$
$$\dot{V}_\mathrm{L} = \mathrm{j}\omega L\dot{I}$$

全体の端子電圧は，

$$\dot{V} = \dot{V}_\mathrm{R} + \dot{V}_\mathrm{L} = \dot{I}R + \mathrm{j}\omega L\dot{I} = (R + \mathrm{j}\omega L)\dot{I}$$

したがって，*RL* 直列回路の合成インピーダンス $\dot{Z}$ [Ω] は以下のように表される.

$$\dot{Z} = \frac{\dot{V}}{\dot{I}} = R + \mathrm{j}\omega L = R + \mathrm{j}X_\mathrm{L}$$

4・12　*RC* 並列回路

抵抗とコンデンサを並列に接続して交流電源につないだ回路は，*RC* 並列回路と呼ばれる.

When the alternating voltage $\dot{V}$ of the angular frequency ω is applied to the *RC* parallel circuit shown in Fig.4・23, the terminal voltage across the parallel branches is equal to $\dot{V}$.

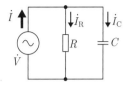

図 4・23　*RC* parallel circuit.

First, we will obtain the combined admittance of the *RC* parallel circuit. When the admittance of R and C are represented by $\dot{Y}_R$ and $\dot{Y}_C$ respectively, then the combined admittance $\dot{Y}$ is obtained by the sum of each admittance as follows.

$$\dot{Y}_R = \frac{1}{R}$$

$$\dot{Y}_C = j\omega C = j\frac{1}{X_C}$$

$$\dot{Y} = \dot{Y}_R + \dot{Y}_C = \frac{1}{R} + j\omega C = \frac{1}{R} + j\frac{1}{X_C}$$

where $\dot{Y}_R$, $\dot{Y}_C$ and $\dot{Y}$ are measured in the unit of siemens (S). X_C is capacitive reactance, $\omega = 2\pi f$ is the angular frequency in rad/s, and f is the frequency in hertz (Hz).

Thus the total current $\dot{I}$ flowing through the circuit can be obtained as follows.

$$\dot{I} = \dot{V}\dot{Y} = \dot{V}(\dot{Y}_R + \dot{Y}_C) = \dot{V}\left(\frac{1}{R} + j\omega C\right) = \dot{V}\left(\frac{1}{R} + j\frac{1}{X_C}\right)$$

The combined admittance of a parallel circuit is the sum of each admittance of the parallel branches. Therefore, by using admittance which is the reciprocal of impedance, the equation is simplified and the calculation for the current of the parallel circuit becomes easy.

4 交流回路

● 角周波数 ω の交流電圧 $\dot{V}$ が，図 $4\cdot23$ に示される RC 並列回路に印加されると，並列枝路の両端の端子電圧は $\dot{V}$ に等しい．

初めに，RC 並列回路の合成アドミタンスを求めることにする．R と C のアドミタンスをそれぞれ $\dot{Y}_\mathrm{R}$ および $\dot{Y}_\mathrm{C}$ で表すと，合成アドミタンス $\dot{Y}$ は次のようにそれぞれのアドミタンスの和で得られる．

$$\dot{Y}_\mathrm{R} = \frac{1}{R}$$

$$\dot{Y}_\mathrm{C} = \mathrm{j}\omega C = \mathrm{j}\frac{1}{X_\mathrm{C}}$$

$$\dot{Y} = \dot{Y}_\mathrm{R} + \dot{Y}_\mathrm{C} = \frac{1}{R} + \mathrm{j}\omega C = \frac{1}{R} + \mathrm{j}\frac{1}{X_\mathrm{C}}$$

ここで，$\dot{Y}_\mathrm{R}$，$\dot{Y}_\mathrm{C}$，$\dot{Y}$ の単位はジーメンス（S）である．X_C は容量性リアクタンス，$\omega = 2\pi f$ は rad/s を単位とする角周波数，f はヘルツ（Hz）を単位とする周波数である．

したがって，回路を流れる全電流 $\dot{I}$ は次のように求められる．

$$\dot{I} = \dot{V}\dot{Y} = \dot{V}(\dot{Y}_\mathrm{R} + \dot{Y}_\mathrm{C}) = \dot{V}\left(\frac{1}{R} + \mathrm{j}\omega C\right) = \dot{V}\left(\frac{1}{R} + \mathrm{j}\frac{1}{X_\mathrm{C}}\right)$$

並列回路の合成アドミタンスは，並列枝路のそれぞれのアドミタンスの和である．したがって，インピーダンスの逆数であるアドミタンスを用いることで，式が単純化され，並列回路の電流の計算が容易になる．

Fig.4·24 shows a phasor diagram for RC parallel circuit. From this Fig., the magnitude I of $\dot{I}$ can be obtained from the following equation using the magnitudes I_R and I_C of $\dot{I}_\mathrm{R}$ and $\dot{I}_\mathrm{C}$ respectively.

$$I = \sqrt{I_\mathrm{R}{}^2 + I_\mathrm{C}{}^2} = \sqrt{\left(\frac{V}{R}\right)^2 + \left(\frac{V}{X_\mathrm{C}}\right)^2} = V\sqrt{\left(\frac{1}{R}\right)^2 + \left(\frac{1}{X_\mathrm{C}}\right)^2}$$
$$= VY$$

where Y and V denote the magnitudes of $\dot{Y}$ and $\dot{V}$ respectively.

From the phasor diagram, we see that $\dot{I}$ leads $\dot{V}$ and the phase difference φ is between 0 and $\pi/2$ rad. The phase difference φ is obtained as follows.

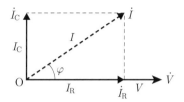

図 4・24　Phasor diagram.

$$\varphi = \tan^{-1}\frac{I_\mathrm{C}}{I_R} = \tan^{-1}\frac{R}{X_\mathrm{C}}$$

The smaller X_C, the more I_C. Therefore, $\dot{I}$ has more phase difference with respect to $\dot{V}$ and the circuit becomes more capacitive.

● 図 4・24 は，RC 並列回路のフェーザ図を表している．この図から，$\dot{I}$ の大きさ I は，$\dot{I}_\mathrm{R}$ と $\dot{I}_\mathrm{C}$ のそれぞれの大きさ I_R と I_C を用いて次式から得られる．

$$I = \sqrt{I_\mathrm{R}{}^2 + I_\mathrm{C}{}^2} = \sqrt{\left(\frac{V}{R}\right)^2 + \left(\frac{V}{X_\mathrm{C}}\right)^2} = V\sqrt{\left(\frac{1}{R}\right)^2 + \left(\frac{1}{X_\mathrm{C}}\right)^2}$$
$$= VY$$

ここで，Y と V はそれぞれ $\dot{Y}$ と $\dot{V}$ の大きさを表す．

　フェーザ図から，$\dot{I}$ は $\dot{V}$ より進み，位相差 φ は 0 と $\pi/2$ rad の間にあることがわかる．位相差 φ は次のように得られる．

$$\varphi = \tan^{-1}\frac{I_\mathrm{C}}{I_R} = \tan^{-1}\frac{R}{X_\mathrm{C}}$$

I_C は X_C が小さくなればなるほど，ますます増加する．したがって，$\dot{I}$ は $\dot{V}$ に対してより大きな位相差をもち，回路はより容量性となる．

4.13　*RLC* 並列回路

　抵抗，コンデンサ，コイルを並列に接続して交流電源につないだ回路は，RLC 並列回路と呼ばれる．

When the alternating voltage $\dot{V}$ of the angular frequency ω is applied to the RLC parallel circuit shown in Fig.4·25, the terminal voltage across the parallel branches is equal to $\dot{V}$.

First, we will obtain the combined admittance of the RLC parallel circuit. When the admittances of R, L, and C are represented by $\dot{Y}_R$, $\dot{Y}_L$, and $\dot{Y}_C$ respectively, then the combined admittance $\dot{Y}$ is obtained by the sum of each admittance as follows.

$$\dot{Y}_R = \frac{1}{R}$$

$$\dot{Y}_L = \frac{1}{j\omega L} = -j\frac{1}{\omega L} = -j\frac{1}{X_L}$$

$$\dot{Y}_C = j\omega C = j\frac{1}{X_C}$$

$$\dot{Y} = \dot{Y}_R + \dot{Y}_L + \dot{Y}_C = \frac{1}{R} - j\frac{1}{\omega L} + j\omega C$$

$$= \frac{1}{R} + j\left(\omega C - \frac{1}{\omega L}\right) = \frac{1}{R} + j\left(\frac{1}{X_C} - \frac{1}{X_L}\right)$$

where $\dot{Y}_R$, $\dot{Y}_L$, $\dot{Y}_C$, and $\dot{Y}$ are measured in the unit of siemens (S). X_L is inductive reactance, X_C is capacitive reactance, $\omega = 2\pi f$ is the angular frequency in rad/s, and f is the frequency in hertz (Hz).

Thus the total current $\dot{I}$ flowing through the circuit can be obtained

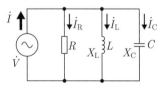

図4·25　RLC parallel circuit.

as follows.

$$\dot{I} = \dot{V}\dot{Y} = \dot{V}(\dot{Y}_R + \dot{Y}_L + \dot{Y}_C) = \dot{V}\left(\frac{1}{R} - j\frac{1}{\omega L} + j\omega C\right)$$

$$= \dot{V}\left\{\frac{1}{R} + j\left(\omega C - \frac{1}{\omega L}\right)\right\} = \dot{V}\left\{\frac{1}{R} + j\left(\frac{1}{X_C} - \frac{1}{X_L}\right)\right\}$$

● 角周波数 ω の交流電圧 $\dot{V}$ が，図 4・25 に示される *RLC* 並列回路に印加されると，並列枝路の両端の端子電圧は $\dot{V}$ に等しい.

初めに，*RLC* 並列回路の合成アドミタンスを求めることにする．R，L，C のアドミタンスをそれぞれ $\dot{Y}_R$，$\dot{Y}_L$，$\dot{Y}_C$ で表すと，合成アドミタンス $\dot{Y}$ は次のようにそれぞれのアドミタンスの和で得られる.

$$\dot{Y}_R = \frac{1}{R}$$

$$\dot{Y}_L = \frac{1}{j\omega L} = -j\frac{1}{\omega L} = -j\frac{1}{X_L}$$

$$\dot{Y}_C = j\omega C = j\frac{1}{X_C}$$

$$\dot{Y} = \dot{Y}_R + \dot{Y}_L + \dot{Y}_C = \frac{1}{R} - j\frac{1}{\omega L} + j\omega C$$

$$= \frac{1}{R} + j\left(\omega C - \frac{1}{\omega L}\right) = \frac{1}{R} + j\left(\frac{1}{X_C} - \frac{1}{X_L}\right)$$

ここで，$\dot{Y}_R$，$\dot{Y}_L$，$\dot{Y}_C$，$\dot{Y}$ の単位はジーメンス（S）である．X_L は誘導性リアクタンス，X_C は容量性リアクタンス，$\omega = 2\pi f$ は rad/s を単位とする角周波数，f はヘルツ（Hz）を単位とする周波数である.

したがって，回路を流れる全電流 $\dot{I}$ は次のように求められる.

$$\dot{I} = \dot{V}\dot{Y} = \dot{V}(\dot{Y}_R + \dot{Y}_L + \dot{Y}_C) = \dot{V}\left(\frac{1}{R} - j\frac{1}{\omega L} + j\omega C\right)$$

$$= \dot{V}\left\{\frac{1}{R} + j\left(\omega C - \frac{1}{\omega L}\right)\right\} = \dot{V}\left\{\frac{1}{R} + j\left(\frac{1}{X_C} - \frac{1}{X_L}\right)\right\}$$

Example In the *RLC* parallel circuit shown in Fig.4·25, when resistance R of 25 Ω, inductive reactance X_L of 100 Ω, and capacitive reactance X_C of 25 Ω are all connected in parallel across 100 V source voltage, calculate the magnitude Y of the combined admittance and the magnitude I of the total current.

例題 図4・25に示す RLC 並列回路で，25 Ωの抵抗 R，100 Ωの誘導性リアクタンス X_L，25 Ωの容量性リアクタンス X_C が100 Vの電源電圧に並列に接続されているとき，合成アドミタンスの大きさ Y と全電流の大きさ I を求めよ．

Solution (答) The magnitude Y of the combined admittance of the circuit is obtained from the following equation.

回路の合成アドミタンスの大きさ Y は次の式から得られる．

$$Y = \sqrt{\left(\frac{1}{R}\right)^2 + \left(\frac{1}{X_C} - \frac{1}{X_L}\right)^2}$$
$$= \sqrt{\left(\frac{1}{25}\right)^2 + \left(\frac{1}{25} - \frac{1}{100}\right)^2} = 0.05\,\text{S}$$

The magnitude I of the total current is obtained from the following equation.

全電流の大きさ I は次の式から得られる．

$$I = VY = 100 \times 0.05 = 5\,\text{A}$$

4.14 直列共振

RLC 直列回路において，周波数を変化させると，ある周波数のとき回路のインピーダンスは最小になる．この現象を直列共振という．

Fig.4・26 shows an RLC series circuit. The impedance $\dot{Z}$ of the circuit is given by

$$\dot{Z} = R + j\omega L + \frac{1}{j\omega C} = R + j\left(\omega L - \frac{1}{\omega C}\right)$$

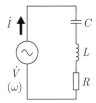

図4・26　RLC series circuit.

where $\dot{Z}$ and R are in ohms, L is in henrys, C is in farads, and ω is the angular frequency of the applied voltage in radians/sec.

The current flowing through the circuit is as follows.

$$\dot{I} = \frac{\dot{V}}{\dot{Z}} = \frac{\dot{V}}{R + j\left(\omega L - \dfrac{1}{\omega C}\right)}$$

At resonance, the imaginary part of $\dot{Z}$ should be equal to zero, i.e.

$$\omega L = \frac{1}{\omega C} \tag{4-21}$$

In this case, the magnitude Z of $\dot{Z}$ is obtained as follows.

$$Z = \sqrt{R^2 + \left(\omega L - \frac{1}{\omega C}\right)^2} = R$$

Therefore, the impedance of the RLC series circuit is at its minimum value. On the other hand, the current is at its maximum value if the voltage is constant as can be seen from the following equation.

$$I = \frac{V}{Z} = \frac{V}{R}$$

In Eq.(4-21), replacing ωL with $2\pi f_0 L$ and $1/\omega C$ with $1/2\pi f_0 C$ gives the following equation.

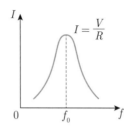

図4・27　series resonance curve.

$$2\pi f_0 L = \frac{1}{2\pi f_0 C}$$

$$\therefore\ f_0 = \frac{1}{2\pi\sqrt{LC}}$$

where frequency f_0 at which resonance occurs is called as resonance frequency in hertz. The resonance frequency f_0 of the series resonance circuit depends only on the inductance L and capacitance C, and does not depend on the resistance R.

Fig.4·27 shows the resonance curve of RLC series circuit. The current is the maximum value at the resonance frequency f_0 and decreases as frequency is changed from the resonance frequency.

● 図4・26は，RLC直列回路を表している．回路のインピーダンス $\dot{Z}$ は以下のように与えられる．

$$\dot{Z} = R + j\omega L + \frac{1}{j\omega C} = R + j\left(\omega L - \frac{1}{\omega C}\right)$$

ここで，$\dot{Z}$ と R はオーム，L はヘンリー，C はファラドを単位とし，ω は印加電圧の角周波数でラジアン／秒を単位とする．
　回路を流れる電流は以下のとおりである．

$$\dot{I} = \frac{\dot{V}}{\dot{Z}} = \frac{\dot{V}}{R + j\left(\omega L - \frac{1}{\omega C}\right)}$$

共振時には，$\dot{Z}$ の虚数部はゼロに等しい．すなわち，

$$\omega L = \frac{1}{\omega C} \tag{4-21}$$

である．

この場合，$\dot{Z}$ の大きさ Z は次のように求められる．

$$Z = \sqrt{R^2 + \left(\omega L - \frac{1}{\omega C}\right)^2} = R$$

したがって，RLC 直列回路のインピーダンスは最小値となる．一方，次式からわかるように，電圧が一定の場合，電流は最大値となる．

$$I = \frac{V}{Z} = \frac{V}{R}$$

(4-21)式で，ωL を $2\pi f_0 L$ に，$1/\omega C$ を $1/2\pi f_0 C$ に置き換えると，次の式が得られる．

$$2\pi f_0 L = \frac{1}{2\pi f_0 C}$$

$$\therefore \ f_0 = \frac{1}{2\pi\sqrt{LC}}$$

ここで，共振が起こる周波数 f_0 は，ヘルツを単位とし，共振周波数と呼ばれる．直列共振回路の共振周波数 f_0 は，インダクタンス L とキャパシタンス C にのみ依存し，抵抗 R には依存しない．

図4・27は，RLC 直列回路の共振曲線を表している．電流は，共振周波数 f_0 で最大値となり，周波数が共振周波数から変化するにつれて減少する．

4.15 交流電力

交流では電圧，電流が常に変化しているので，その積として与えられる電力もまた変化する．

There are three types of powers in AC circuit: active power(P), reactive power(Q), and apparent power(S).

Active power is the power consumed by resistor and is measured in the unit of watts(W). For example, in the AC circuit in Fig.4・28(a), the active power can be expressed by multiplying of the square of the current by the resistance. Also, to obtain the value of active power as the product of voltage and current, it is necessary to multiply this product by the cosine of the phase difference θ between the voltage and the current.

The active power is expressed as follows.

$$P = I^2 R = VI \cos \theta \ [\text{W}]$$

where V and I are voltage and current respectively and are in effective values to obtain the active power in watts.

In the AC circuit shown in Fig. 4.28(a), the current I flowing through the circuit can be obtained as follows with reference to Fig.4·28(b).

$$I = \frac{V}{\sqrt{R^2 + X_L^2}} = \frac{200}{\sqrt{200^2 + 346^2}} = \frac{200}{400} = 0.5 \ \text{A}$$
$$\therefore \quad P = I^2 R = 0.5^2 \times 200 = 50 \ \text{W} \qquad (4\text{-}22)$$

Moreover, the applied voltage is 200 V. Since the phase difference between the voltage and the current is $\pi/3$ rad, the active power is obtained as follows.

$$P = VI \cos \theta = 200 \times 0.5 \times \cos \frac{\pi}{3} = 50 \ \text{W} \qquad (4\text{-}23)$$

Therefore, the active power is the same in Eqs.(4-22) and (4-23).

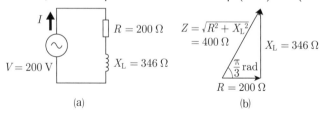

図 4・28　RL series circuit.

Both formulas can be used to calculate the active power.

This cos θ is called power factor, which represents the ratio of active power to apparent power. As the power factor is closer to one, the power is used more effectively.

● 交流回路には，有効電力（P），無効電力（Q），皮相電力（S）の3種類の電力がある．

有効電力は，抵抗によって消費される電力で，ワット（W）の単位で測定される．例えば，図4・28(a)の交流回路において，有効電力は，電流の二乗に抵抗を掛けて表すことができる．また，有効電力の値を電圧と電流の積として求めるためには，この積に電圧と電流の位相差θの余弦を掛ける必要がある．

有効電力は，次のように表される．

$$P = I^2 R = VI \cos \theta \ [\text{W}]$$

ここで，V と I はそれぞれ電圧と電流で，有効電力をワットを単位として得ることから実効値である．図4・28(a)の交流回路において，回路を流れる電流Iは，図4・28(b)を参考にして次のように得ることができる．

$$I = \frac{V}{\sqrt{R^2 + X_\mathrm{L}{}^2}} = \frac{200}{\sqrt{200^2 + 346^2}} = \frac{200}{400} = 0.5 \ \text{A}$$

$$\therefore \ \ P = I^2 R = 0.5^2 \times 200 = 50 \ \text{W} \tag{4-22}$$

さらに，この回路において，印加電圧は200Vである．電圧と電流の位相差は$\pi/3$ radなので，有効電力は次のように得られる．

$$P = VI \cos \theta = 200 \times 0.5 \times \cos \frac{\pi}{3} = 50 \ \text{W} \tag{4-23}$$

したがって，有効電力は(4-22)式と(4-23)式で同じである．どちらの公式も有効電力の計算に使用できる．

この cos θ は力率と呼ばれ，有効電力と皮相電力の比を表す．力率は1に近いほど，電力がより活用される．

Reactive power is energy stored in coils and capacitors, which is not actually consumed. In the AC circuit in Fig. 4・28(a), the reactive power is the value of the inductive reactance multiplied by the square of the current flowing through the circuit. Also, to obtain the value of reactive power as the product of voltage and current, it is necessary to multiply this product by the sine of the phase difference θ between the voltage and the current. Namely,

$$Q = I^2 X_L = VI \sin \theta \text{ [var]}$$

The reactive power is measured in the unit of Volt-Amps-Reactive(var). The reactive power in the AC circuit in Fig. 4·28(a) is obtained as follows.

$$Q = I^2 X_L = 0.5^2 \times 346 = 86.5 \text{ var}$$

$$Q = VI \sin \theta = 200 \times 0.5 \times \sin \frac{\pi}{3} \fallingdotseq 86.6 \text{ var}$$

● 無効電力は，コイルやコンデンサに蓄えられたエネルギーで，実際には消費されない．図4・28(a)の交流回路において，無効電力は，誘導性リアクタンスに回路を流れる電流の2乗を掛けた値である．また，無効電力の値を電圧と電流の積として求めるには，この積に電圧と電流の位相差θの正弦を掛ける必要がある．すなわち，

$$Q = I^2 X_L = VI \sin \theta \text{ [var]}$$

無効電力は，バール（var）の単位で測定される．図4・28(a)の交流回路における無効電力は，次のようにして得られる．

$$Q = I^2 X_L = 0.5^2 \times 346 = 86.5 \text{ var}$$

$$Q = VI \sin \theta = 200 \times 0.5 \times \sin \frac{\pi}{3} \fallingdotseq 86.6 \text{ var}$$

Apparent power is the power supplied to the circuit and measured in the unit of Volt-Amps(V·A). The apparent power is the vector sum of the active and reactive power, and it is the product of the voltage and the current. Namely,

$$S = VI \text{ [V·A]}$$

The apparent power in the AC circuit in Fig.4·28(a) is obtained as follows.

$$S = VI = 200 \times 0.5 = 100 \text{ V·A}$$

Since apparent power is useful for calculating how much current flows at rated voltage, it is generally used to represent the capacity of a power equipment.

● 皮相電力は，回路に供給される電力のことで，ボルトアンペア（V·A）の単位で測定される．皮相電力は，有効電力と無効電力のベクトル和で，電圧と電流の積である．すなわち，

$$S = VI$$

図4·28(a)の交流回路における皮相電力は，次のようにして得られる．

$$S = VI = 200 \times 0.5 = 100 \text{ V·A}$$

皮相電力は，定格電圧においてどのくらいの電流が流れるかを計算するのに役立つので，一般に電気機器の容量を表すのに用いられる．

The right triangle in Fig.4·29 shows the phase relationship between apparent power, active power, and reactive power.

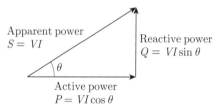

Apparent power
$S = VI$

Reactive power
$Q = VI \sin \theta$

θ

Active power
$P = VI \cos \theta$

図4·29 Power triangle.

● 第4.29図の直角三角形は，皮相電力，有効電力，無効電力の位相関係を表している．

4.16 交流ブリッジ

直流回路でのホイートストンブリッジと同様に，交流回路では抵抗だけでなく，コイルやコンデンサからなるインピーダンスの測定に応用できる交流ブリッジがある．

As shown in Fig.4·30, AC bridge consists of four impedance arms

4 交流回路

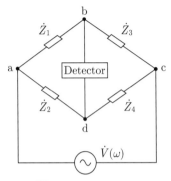

図 4・30 AC bridge.

$\dot{Z}_1$, $\dot{Z}_2$, $\dot{Z}_3$, and $\dot{Z}_4$, an AC voltage source, and a detector.

AC bridges are used to measure unknown resistance, inductance, and capacitance values. The AC bridges are very convenient and give the accurate result of the measurement.

When the current flowing through the detector is zero, the AC bridge is in a balanced condition. In this balanced condition, the following equation is satisfied.

$$\dot{Z}_1 \dot{Z}_4 = \dot{Z}_2 \dot{Z}_3$$

● 図4・30に示すように，交流ブリッジは，四つのインピーダンス辺 $\dot{Z}_1$，$\dot{Z}_2$，$\dot{Z}_3$ および $\dot{Z}_4$，交流電圧源，それに検出器で構成される．

交流ブリッジは，未知の抵抗，インダクタンス，およびキャパシタンスの値を測定するために使用される．交流ブリッジは非常に便利で，正確な測定結果をもたらす．

検出器を流れる電流がゼロのとき，交流ブリッジは平衡状態にある．平衡状態では，次式が成り立つ．

$$\dot{Z}_1 \dot{Z}_4 = \dot{Z}_2 \dot{Z}_3$$

Fig.4·31 shows Maxwell bridge. This bridge is used to measure the resistance R_x and inductance L_x of a coil. The arms ad and bc consist of resistances R_2 and resistance R_3 respectively. The arm ab consists of a parallel combination of resistance R_1 and capacitor C_1. The arm cd consists of a series combination of resistance R_x and inductor L_x.

Let $\dot{Z}_1$, $\dot{Z}_2$, $\dot{Z}_3$ and $\dot{Z}_x$ are the impedances of arms ab, da, bc, and cd respectively. Impedance $\dot{Z}_1$ is the reciprocal of admittance $\dot{Y}_1$.

The values of these admittance and impedances will be

$$\dot{Y}_1 = \frac{1}{\dot{Z}_1} = \frac{1}{R_1} + j\omega C_1 = \frac{1 + j\omega C_1 R_1}{R_1}$$
$$\dot{Z}_2 = R_2$$
$$\dot{Z}_3 = R_3$$
$$\dot{Z}_x = R_x + j\omega L_x$$

At balance we get

$$\dot{Z}_1 \dot{Z}_x = \dot{Z}_2 \dot{Z}_3$$

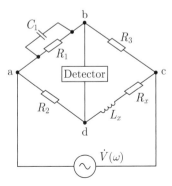

図4·31 Maxwell bridge.

$$\therefore\ \dot{Z}_x = \frac{\dot{Z}_2 \dot{Z}_3}{\dot{Z}_1} = \dot{Y}_1 \dot{Z}_2 \dot{Z}_3$$

Therefore,

$$R_x + j\omega L_x = \frac{1 + j\omega C_1 R_1}{R_1} R_2 R_3$$

$$= \frac{R_2 R_3 + j\omega C_1 R_1 R_2 R_3}{R_1}$$

$$= \frac{R_2 R_3}{R_1} + j\omega C_1 R_2 R_3 \tag{4-24}$$

The following equations are obtained by comparing the real and imaginary parts of both sides of Eq.(4-24)

$$R_x = \frac{R_2 R_3}{R_1} \tag{4-25}$$

$$L_x = C_1 R_2 R_3 \tag{4-26}$$

Hence, the unknown resistance R_x and unknown inductance L_x can be determined.

● 図4・31はマクスウェルブリッジを示している．このブリッジは，コイルの抵抗 R_x とインダクタンス L_x を測定するために用いられる．辺da と bc は，それぞれ抵抗 R_2 と R_3 で構成されている．辺ab は，抵抗 R_1 とコンデンサ C_1 の並列接続で構成されている．辺cd は，抵抗 R_x とインダクタンス L_x の直列接続である．

$\dot{Z}_1$, $\dot{Z}_2$, $\dot{Z}_3$, $\dot{Z}_x$ をそれぞれ辺ab，da，bc，cd のインピーダンスとする．インピーダンス $\dot{Z}_1$ は，アドミタンス $\dot{Y}_1$ の逆数である．

これらのアドミタンスとインピーダンスの値は

$$\dot{Y}_1 = \frac{1}{\dot{Z}_1} = \frac{1}{R_1} + j\omega C_1 = \frac{1 + j\omega C_1 R_1}{R_1}$$
$$\dot{Z}_2 = R_2$$

$$\dot{Z}_3 = R_3$$
$$\dot{Z}_x = R_x + j\omega L_x$$

平衡時には

$$\dot{Z}_1 \dot{Z}_x = \dot{Z}_2 \dot{Z}_3$$

$$\therefore \quad \dot{Z}_x = \frac{\dot{Z}_2 \dot{Z}_3}{\dot{Z}_1} = \dot{Y}_1 \dot{Z}_2 \dot{Z}_3$$

となる．それゆえ，

$$R_x + j\omega L_x = \frac{1 + j\omega C_1 R_1}{R_1} R_2 R_3$$

$$= \frac{R_2 R_3 + j\omega C_1 R_1 R_2 R_3}{R_1}$$

$$= \frac{R_2 R_3}{R_1} + j\omega C_1 R_2 R_3 \tag{4-24}$$

(4-24)式の両辺の実部と虚部を比較することで次の式が得られる．

$$R_x = \frac{R_2 R_3}{R_1} \tag{4-25}$$

$$L_x = C_1 R_2 R_3 \tag{4-26}$$

したがって，未知抵抗 R_x および未知インダクタンス L_x を求めることができる．

Example In Maxwell bridge shown in Fig.4·31, let $R_1 = 1\,\text{k}\Omega$, $R_2 = 600\,\Omega$, $R_3 = 400\,\Omega$, and $C_1 = 0.5\,\mu\text{F}$. Find the values of the unknown resistance R_x and inductance L_x at balance.

例題 図4·31に示すマクスウェルブリッジにおいて，$R_1 = 1\,\text{k}\Omega$，$R_2 = 600\,\Omega$，$R_3 = 400\,\Omega$，$C_1 = 0.5\,\mu\text{F}$ とする．平衡時における未知抵抗 R_x と未知インダクタンス L_x の値を求めよ．

Solution(答) From Eqs.(4-25) and (4-26).

(4-25)式と(4-26)式より，

$$R_x = \frac{R_2 R_3}{R_1} = \frac{600 \times 400}{1\,000} = 240\,\Omega$$

$$L_x = C_1 R_2 R_3 = 0.5 \times 10^{-6} \times 600 \times 400$$
$$= 0.12 \text{ H} = 120 \text{ mH}$$

4.17　三相交流の大きさ

　三つの電圧の大きさが等しく，位相が互いに120°異なる交流を三相交流という．

　Fig.4·32 shows the principle of a three phase alternator. As shown in this Fig., three coils, which are arranged to make 120° each other, are rotated counterclockwise in a magnetic field. As a result, three independent alternating voltages will be induced across the coils. They have the same amplitude and freguency, and are out of phase with each other by 120°. The waveforms are shown in Fig.4·33. Let v_a, v_b, and v_c be the three independent voltages induced in coils a-a',b-b', and c-c' respectively. Assuming that v_a is the reference, the instantaneous values of the electromotive force v_a, v_b, and v_c [V] are expressed as follows.

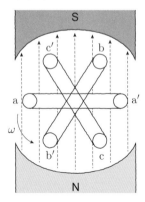

図4·32　Generation of three phase AC.

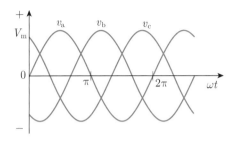

図4・33　Waveforms of three phase voltages.

$v_a = V_m \sin \omega t$

$v_b = V_m \sin(\omega t - 120°)$

$v_c = V_m \sin(\omega t - 240°)$

where V_m [V] is the maximum value of the electromotive force of each coil, and ω [rad/s] is the angular velocity of the coil.

　The phasor diagram of these voltages can be shown as in Fig.4·34. As phasors rotate in counterclockwise, we can say that v_b lags v_a by 120° and v_c lags v_b by 120°. That is, the voltage induced in each coil becomes maximum in order of v_a, v_b, and v_c with reference to v_a. This sequence is called phase sequence.

　If we add three voltages vectorially, it can be observed that the sum

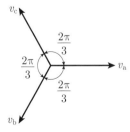

図4・34　Three phase phasor diagram.

of the three voltages at instant is zero. This can be expressed as follows using equations.

$$v_a + v_b + v_c$$
$$= V_m \sin \omega t + V_m \sin(\omega t - 120°) + V_m \sin(\omega t - 240°)$$
$$= V_m (\sin \omega t + \sin \omega t \cos 120° - \cos \omega t \sin 120°$$
$$\qquad + \sin \omega t \cos 240° - \cos \omega t \sin 240°)$$
$$= V_m \left\{ \sin \omega t - \frac{1}{2} \sin \omega t - \frac{\sqrt{3}}{2} \cos \omega t - \frac{1}{2} \sin \omega t \right.$$
$$\qquad \left. + \frac{\sqrt{3}}{2} \cos \omega t \right\}$$
$$= 0$$

● 図4・32に三相交流発電機の原理図を示す．この図に示すように，互いに120°をなすように配置された三つのコイルが磁界中で反時計回りに回転する．その結果，三つの独立した交流電圧がコイルの両端に誘導される．それらは同じ大きさと周波数をもち，互いに120°位相がずれている．その波形を図4・33に示す．

ここで，v_a, v_b, v_cをそれぞれコイル a-a′, b-b′, c-c′ に誘起される三つの独立した電圧とする．v_aを基準とすれば，起電力の瞬時値 v_a, v_b, v_c [V] はそれぞれ次のように表される．

$$v_a = V_m \sin \omega t$$
$$v_b = V_m \sin(\omega t - 120°)$$
$$v_c = V_m \sin(\omega t - 240°)$$

ここで，V_m [V] は各コイルの起電力の最大値，ω [rad/s] はコイルの角速度である．

これらの電圧のフェーザ図は，図4・34のように表すことができる．フェーザが反時計回りに回転すると，v_bはv_aより120°遅れ，v_cはv_bより120°遅れる．すなわち，各コイルに誘起される電圧はv_aを基準としてv_a，v_b, v_cの順に最大となる．この順序は相順と呼ばれる．

ベクトル的に三つの電圧を加算すると，瞬時のそれらの電圧の合計はゼロであることがわかる．これは式を用いて次のように表せる．

$$v_a + v_b + v_c$$
$$= V_m \sin \omega t + V_m \sin(\omega t - 120°) + V_m \sin(\omega t - 240°)$$
$$= V_m (\sin \omega t + \sin \omega t \cos 120° - \cos \omega t \sin 120°$$
$$\quad + \sin \omega t \cos 240° - \cos \omega t \sin 240°)$$
$$= V_m \left\{ \sin \omega t - \frac{1}{2} \sin \omega t - \frac{\sqrt{3}}{2} \cos \omega t - \frac{1}{2} \sin \omega t \right.$$
$$\left. \quad + \frac{\sqrt{3}}{2} \cos \omega t \right\}$$
$$= 0$$

4.18　平衡三相回路（△-△結線）

　対称起電力で負荷が同じインピーダンスをもつ平衡三相回路において，起電力と負荷がともに△結線である回路を平衡 △-△ 三相回路という．

　Fig.4·35 shows a balanced △-△ three phase circuit in which both the electromotive force and the load are △ connected. The electromotive forces $\dot{V}_a$, $\dot{V}_b$, and $\dot{V}_c$ are called phase voltages. Assuming that $\dot{V}_a$ is a reference point in phase sequence, each phase voltage is as follows.

$$\dot{V}_a = V_m \angle 0°$$
$$\dot{V}_b = V_m \angle -120°$$
$$\dot{V}_c = V_m \angle -240°$$

where V_m is the magnitude of each phase voltage.

　For a △-△ connection, the phase voltage and the line voltage have the following relationship.

$$\dot{V}_a = \dot{V}_{ab}$$
$$\dot{V}_b = \dot{V}_{bc}$$

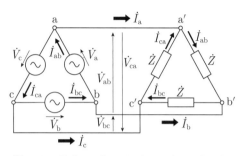

図4・35 Balanced △-△ three phase circuit.

$$\dot{V_c} = \dot{V}_{ca}$$

That is, since the magnitude V_m of the phase voltage is equal to the magnitude V_L of the line voltage, the following relationship holds.

$$V_m = V_L$$

The △-△ circuit can be considered as a circuit in which each of the three phases is independent. Therefore, the phase currents $\dot{I}_{ab}$, $\dot{I}_{bc}$, and $\dot{I}_{ca}$ can be expressed as follows using the phase voltages $\dot{V}_a$, $\dot{V}_b$, and $\dot{V}_c$ and the impedance $\dot{Z}$.

$$\dot{I}_{ab} = \frac{\dot{V}_a}{\dot{Z}}, \quad \dot{I}_{bc} = \frac{\dot{V}_b}{\dot{Z}}, \quad \dot{I}_{ca} = \frac{\dot{V}_c}{\dot{Z}}$$

Since line voltages are balanced, phase currents are also balanced. Assuming that $\dot{I}_{ab}$ is a reference of the current, each phase current can be expressed as follows.

$$\left.\begin{array}{l} \dot{I}_{ab} = I_m \angle\ 0° \\ \dot{I}_{bc} = I_m \angle\ -120° \\ \dot{I}_{ca} = I_m \angle\ -240° \end{array}\right\} \tag{4-27}$$

where I_m is the magnitude of each phase current.

The line currents $\dot{I}_a$, $\dot{I}_b$, and $\dot{I}_c$ are obtained from the phase currents by applying Kirchhoff's current law at nodes a′, b′, and c′ respectively.

$$\left.\begin{array}{l} \dot{I}_a = \dot{I}_{ab} - \dot{I}_{ca} \\ \dot{I}_b = \dot{I}_{bc} - \dot{I}_{ab} \\ \dot{I}_c = \dot{I}_{ca} - \dot{I}_{bc} \end{array}\right\} \tag{4-28}$$

The following relationship is obtained from Eqs.(4-27) and (4-28).

$$\begin{aligned} \dot{I}_a &= I_m(\cos 0° + j\sin 0°) \\ &\quad - I_m\{\cos(-240°) + j\sin(-240°)\} \\ &= I_m - I_m\left(-\frac{1}{2} + j\frac{\sqrt{3}}{2}\right) = I_m\left(\frac{3}{2} - j\frac{\sqrt{3}}{2}\right) \\ &= \sqrt{3}I_m\angle -30° \end{aligned}$$

$$\begin{aligned} \dot{I}_b &= I_m\{\cos(-120°) + j\sin(-120°)\} \\ &\quad - I_m(\cos 0° + j\sin 0°) \\ &= I_m\left(-\frac{1}{2} - j\frac{\sqrt{3}}{2}\right) - I_m = I_m\left(-\frac{3}{2} - j\frac{\sqrt{3}}{2}\right) \\ &= \sqrt{3}I_m\angle -150° \end{aligned}$$

$$\begin{aligned} \dot{I}_c &= I_m\{\cos(-240°) + j\sin(-240°)\} \\ &\quad - I_m\{\cos(-120°) + j\sin(-120°)\} \\ &= I_m\left(-\frac{1}{2} + j\frac{\sqrt{3}}{2}\right) - I_m\left(-\frac{1}{2} - j\frac{\sqrt{3}}{2}\right) \\ &= j\sqrt{3}I_m = I\sqrt{3}I_m\angle -270° \end{aligned}$$

That is, each line current lags the corresponding phase current by 30° and the magnitude of the line current is $\sqrt{3}$ times the magnitude of the phase current.

Fig.4·36 shows the phasor diagram for △-△ connection. From this Fig., it is clear that each line current is the difference of phase

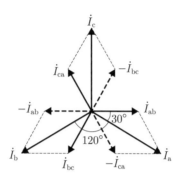

図4・36　Phasor diagram of a △-△ connection circuit.

currents of two phases concerned. For example, the line current $\dot{I}_b$ will be equal to the difference of phase currents $\dot{I}_{bc}$ and $\dot{I}_{ab}$.

● 図4・35は，起電力と負荷の両方が△結線された平衡△-△三相回路を示している．起電力$\dot{V}_a$，$\dot{V}_b$，$\dot{V}_c$は相電圧と呼ばれる．

$\dot{V}_a$を相順の基準点とすると，各相電圧は次のようになる．

$$\dot{V}_a = V_m \angle 0°$$
$$\dot{V}_b = V_m \angle -120°$$
$$\dot{V}_c = V_m \angle -240°$$

ここで，V_mは各相電圧の大きさである．

△-△結線の場合，相電圧と線間電圧には次の関係がある．

$$\dot{V}_a = \dot{V}_{ab}$$
$$\dot{V}_b = \dot{V}_{bc}$$
$$\dot{V}_c = \dot{V}_{ca}$$

すなわち，相電圧の大きさV_mは，線間電圧の大きさV_Lに等しいので，次の関係が成り立つ．

$$V_m = V_L$$

△-△回路は，三相の各相が独立した回路として考えることができる．したがって，相電流$\dot{I}_{ab}$，$\dot{I}_{bc}$，$\dot{I}_{ca}$は，相電圧$\dot{V}_a$，$\dot{V}_b$，$\dot{V}_c$とインピーダンス$\dot{Z}$を用いて次のように表すことができる．

$$\dot{I}_{ab} = \frac{\dot{V}_a}{\dot{Z}}, \quad \dot{I}_{bc} = \frac{\dot{V}_b}{\dot{Z}}, \quad \dot{I}_{ca} = \frac{\dot{V}_c}{\dot{Z}}$$

線間電圧は平衡しているので，相電流も平衡している．$\dot{I}_{ab}$ を電流の基準にとると，各相電流は次のように表すことができる．

$$\left.\begin{array}{l} \dot{I}_{ab} = I_m \angle 0° \\ \dot{I}_{bc} = I_m \angle -120° \\ \dot{I}_{ca} = I_m \angle -240° \end{array}\right\} \tag{4-27}$$

ここで，I_m は相電流の大きさである．

線電流 $\dot{I}_a$，$\dot{I}_b$，および $\dot{I}_c$ は，それぞれ a′点，b′点，c′点でキルヒホッフの電流則を適用することで相電流から求められる．

$$\left.\begin{array}{l} \dot{I}_a = \dot{I}_{ab} - \dot{I}_{ca} \\ \dot{I}_b = \dot{I}_{bc} - \dot{I}_{ab} \\ \dot{I}_c = \dot{I}_{ca} - \dot{I}_{bc} \end{array}\right\} \tag{4-28}$$

(4-27)式と(4-28)式から次の関係が得られる．

$$\begin{aligned} \dot{I}_a &= I_m(\cos 0° + \mathrm{j}\sin 0°) \\ &\quad - I_m\{\cos(-240°) + \mathrm{j}\sin(-240°)\} \\ &= I_m - I_m\left(-\frac{1}{2} + \mathrm{j}\frac{\sqrt{3}}{2}\right) = I_m\left(\frac{3}{2} - \mathrm{j}\frac{\sqrt{3}}{2}\right) \\ &= \sqrt{3}I_m \angle -30° \end{aligned}$$

$$\begin{aligned} \dot{I}_b &= I_m\{\cos(-120°) + \mathrm{j}\sin(-120°)\} \\ &\quad - I_m(\cos 0° + \mathrm{j}\sin 0°) \\ &= I_m\left(-\frac{1}{2} - \mathrm{j}\frac{\sqrt{3}}{2}\right) - I_m = I_m\left(-\frac{3}{2} - \mathrm{j}\frac{\sqrt{3}}{2}\right) \\ &= \sqrt{3}I_m \angle -150° \end{aligned}$$

$$\begin{aligned} \dot{I}_c &= I_m\{\cos(-240°) + \mathrm{j}\sin(-240°)\} \\ &\quad - I_m\{\cos(-120°) + \mathrm{j}\sin(-120°)\} \\ &= I_m\left(-\frac{1}{2} + \mathrm{j}\frac{\sqrt{3}}{2}\right) - I_m\left(-\frac{1}{2} - \mathrm{j}\frac{\sqrt{3}}{2}\right) \\ &= \mathrm{j}\sqrt{3}I_m = I\sqrt{3}I_m \angle -270° \end{aligned}$$

すなわち，各線電流は，対応する相電流に対して30°遅れ，線電流の大きさは，相電流の大きさの $\sqrt{3}$ 倍である．

図4・36に △-△ 結線のフェーザ図を示す．この図から，各線電流は関係する二つの相の相電流の差であることが明らかである．例えば，線電流 $\dot{I}_b$ は，相電流 $\dot{I}_{bc}$ と $\dot{I}_{ab}$ の差に等しい．

4.19 ひずみ波

一定の周期で変化するが，正弦波でない交流を一般に非正弦波交流，またはひずみ波という．

Non-sinusoidal wave AC is a waveform different from a sine or cosine wave and is also called a distorted wave. The distorted wave is encountered in many areas of engineering. Typical examples include a square wave, a triangular wave, and a saw-tooth wave.

Fig.4·37 shows a square wave. The square wave is a non-sinusoidal periodic waveform whose amplitude changes from maximum to minimum instantly, where the pulse width is equal to one-half the period. In the digital circuit based on binary logic, values 0 and 1 are associated with two different voltage levels. For example, logic 0 is represented by the low voltage level and logic 1 by the high voltage level. The square wave is used to represent these two voltage levels in the digital circuit because its condition can be changed easily.

The square wave is also used for various types of audio signal as well as for clock and other digital signal in electronic and digital circuits.

● 非正弦波交流は，正弦波または余弦波とは異なる波形のことで，ひずみ波とも呼ばれる．ひずみ波は，工学の多くの分野で見受けられる．代表的

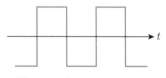

図4·37 Square waveform.

な例として，方形波，三角波それにのこぎり波がある．

　図4・37に方形波を示す．方形波は，振幅が瞬時に最大値から最小値に変化する非正弦波の周期的な波形で，パルス幅は周期の半分に等しい．2値論理に基づくディジタル回路では，値0と1は二つの電圧レベルに関連付けられる．例えば，論理0は低い電圧レベルで，論理1は高い電圧レベルで表される．方形波は，状態を容易に変更できるため，ディジタル回路でこれらの二つの電圧レベルを表すために使用される．

　方形波は，電子回路やディジタル回路のクロックやその他のディジタル信号だけでなく，さまざまな種類のオーディオ信号にも使用される．

Fig.4・38 shows a triangular waveform. Since the slope of the rising part and the slope of the falling part are equal in the triangular wave, the rise time and fall time are equal.

The triangular wave is used as a signal source for creating a PWM wave using a comparator. Moreover, in order to use a triangular wave for noise removal, a filter circuit that converts a square wave into a triangular wave is often used.

図4・38　Triangular waveform.

● 　図4・38に三角波を示す．三角波では，立上り部分の傾きと立下り部分の傾きが等しいため，立上り時間と立下り時間は等しくなる．

　三角波は，比較器を用いてパルス幅変調波を生成するための信号源として使用される．また，ノイズ除去に三角波を用いるために，方形波を三角波に変換するフィルタ回路がしばしば用いられる．

Fig.4・39 shows a saw-tooth waveform. As you can see from this Fig., the saw-tooth wave rises slowly and linearly on each cycle and then falls straight down to its starting value.

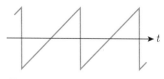

図 4・39　Saw-tooth waveform.

The saw-tooth waveform is used to trigger the operations of an electronic circuit. In a television set and an oscilloscope, it is used to create the image by sweeping an electron beam across the screen.

● 図4・39にのこぎり波を示す．この図からわかるように，のこぎり波は各サイクルでゆっくりと直線的に上昇し，その後開始値まで真っすぐに下降する．

のこぎり波は，電子回路の動作をトリガするために使用される．テレビやオシロスコープでは，画面全体に電子ビームを掃引して画像を作成するために使用される．

4.20　フーリエ級数展開

一般に，任意のひずみ波の波形は，複数の正弦波を重ね合わせたものと考えることができる．これらの正弦波の和はフーリエ級数に展開したものである．

In 1822, French mathematician Joseph Fourier introduced the idea that any wave can be represented as an infinite sum of different types of sine waves. The sum of these sine waves is called the Fourier series, and distorted waves are handled using the Fourier series. The voltage $v(t)$ of the distorted wave shown in Fig.4・40 can be expressed by the following Fourier series.

$$v(t) = a_0 + a_1 \sin \omega t + a_2 \sin 2\omega t + a_3 \sin 3\omega t + \cdots$$
$$+ b_1 \cos \omega t + b_2 \cos 2\omega t + b_3 \cos 3\omega t + \cdots$$

$$= a_0 + \sum_{n=1}^{\infty} a_n \sin n\omega t + \sum_{n=1}^{\infty} b_n \cos n\omega t$$

$$= a_0 + \sum_{n=1}^{\infty} \left(a_n \sin n\omega t + b_n \cos n\omega t \right) \qquad (4\text{-}29)$$

where let $A_n = \sqrt{a_n{}^2 + b_n{}^2}$, $\theta_n = \tan^{-1} \dfrac{b_n}{a_n}$, then $v(t)$ can be written as follows.

$$v(t) = a_0 + \sum_{n=1}^{\infty} A_n \sin(n\omega t + \theta_n) \qquad (4\text{-}30)$$

The first term a_0 in Eq.(4-30) represents the average value of the voltage of the distorted wave. If the area above time axis is equal to

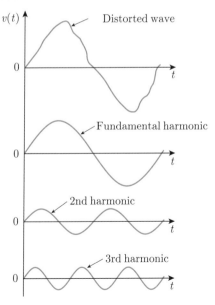

図4・40 Distorted wave and harmonics.

the area below time axis in one period, a_0 becomes 0.

As shown in Fig.4·40, a sine wave having the same frequency as the distorted wave is called the fundamental harmonic among the sine wave ACs that constitutes the distorted wave. The harmonic waves having the frequencies twice and three times as the fundamental harmonic are called the second and third harmonics respectively. In general, a higher harmonic with a frequency n times as the fundamental harmonic is called the nth harmonic.

Therefore, a distorted wave is generally as follows.

Distorted wave = Average value (DC component)

　　　　　　　　　　+ Fundamental harmonic + Higher harmonic

The Fourier series in Eq.(4-29) is transformed as follows using Euler's formula.

Euler's formula states that

$$e^{jn\omega t} = \cos n\omega t + j \sin n\omega t \qquad (4\text{-}31)$$

From Eq.(4-31)

$$\cos n\omega t = \frac{e^{jn\omega t} + e^{-jn\omega t}}{2}, \quad \sin n\omega t = \frac{e^{jn\omega t} - e^{-jn\omega t}}{2j}$$

$$\therefore \ v(t) = a_0 + \sum_{n=1}^{\infty} \left(a_n \sin n\omega t + b_n \cos n\omega t \right)$$

$$= a_0 + \sum_{n=1}^{\infty} \left(\frac{b_n - ja_n}{2} e^{jn\omega t} + \frac{b_n + ja_n}{2} e^{-jn\omega t} \right)$$

Finally, we can summarize as follows by applying the Fourier series to each of square wave, triangular wave, and saw-tooth wave.

A square waveform is composed of the fundamental harmonic and all odd harmonics. A triangular waveform is also composed of the fundamental harmonic and all odd harmonics, but even-numbered terms in the Fourier series expansion have a negative sign. A sawtooth waveform consists of both even and odd harmonics.

● 1822年，フランスの数学者ジョセフ・フーリエは，任意の波を異なる種類の正弦波の無限の和として表すことができるという考えを紹介した．これらの正弦波の和は，フーリエ級数と呼ばれ，ひずみ波はフーリエ級数を利用して処理される．第4.40図に示すひずみ波の電圧 $v(t)$ は，以下のフーリエ級数で表すことができる．

$$v(t) = a_0 + a_1 \sin \omega t + a_2 \sin 2\omega t + a_3 \sin 3\omega t + \cdots$$
$$+ b_1 \cos \omega t + b_2 \cos 2\omega t + b_3 \cos 3\omega t + \cdots$$

$$= a_0 + \sum_{n=1}^{\infty} a_n \sin n\omega t + \sum_{n=1}^{\infty} b_n \cos n\omega t$$

$$= a_0 + \sum_{n=1}^{\infty} \left(a_n \sin n\omega t + b_n \cos n\omega t \right) \tag{4-29}$$

ここで，$A_n = \sqrt{a_n{}^2 + b_n{}^2}$，$\theta_n = \tan^{-1} \dfrac{b_n}{a_n}$ とおくと $v(t)$ は次のように表すことができる．

$$v(t) = a_0 + \sum_{n=1}^{\infty} A_n \sin(n\omega t + \theta_n) \tag{4-30}$$

(4-30)式の第1項 a_0 は，ひずみ波の平均値を表す．1周期で時間軸より上の領域が時間軸より下の領域と等しい場合，$a_0 = 0$ になる．

図4・40に示すように，ひずみ波を構成する正弦波交流のうち，ひずみ波と同じ周波数の正弦波は基本波と呼ばれる．基本数の2倍と3倍の周波数をもつ高調波は，それぞれ第2調波，第3調波と呼ばれる．一般に，基本波の n 倍の周波数をもつ調波は，第 n 調波と呼ばれる．

したがって，ひずみ波は，一般に次のようになる．

　　　ひずみ波 ＝ 平均値(直流分) ＋ 基本波＋高調波

(4-29)式のフーリエ級数は，オイラーの公式を用いて次のように変換される．

　　オイラーの公式は

$$\mathrm{e}^{\mathrm{j}n\omega t} = \cos n\omega t + \mathrm{j} \sin n\omega t \tag{4-31}$$

である．

　　(4-31)式から

$$\cos n\omega t = \frac{\mathrm{e}^{\mathrm{j}n\omega t} + \mathrm{e}^{-\mathrm{j}n\omega t}}{2}, \quad \sin n\omega t = \frac{\mathrm{e}^{\mathrm{j}n\omega t} - \mathrm{e}^{-\mathrm{j}n\omega t}}{2\mathrm{j}}$$

$$\therefore \ v(t) = a_0 + \sum_{n=1}^{\infty} \left(a_n \sin n\omega t + b_n \cos n\omega t \right)$$

$$= a_0 + \sum_{n=1}^{\infty} \left(\frac{b_n - \mathrm{j}a_n}{2} \mathrm{e}^{\mathrm{j}n\omega t} + \frac{b_n + \mathrm{j}a_n}{2} \mathrm{e}^{-\mathrm{j}n\omega t} \right)$$

最後に，方形波，三角波，のこぎり波のそれぞれにフーリエ級数を適用することにより，次のように要約できる．

方形波は，基本波とすべての奇数調波で構成される．三角波も基本波とすべての奇数調波で構成されるが，フーリエ級数展開の偶数番目の項には負の符号が付く．のこぎり波は，偶数調波と奇数調波で構成される．

英 和 索 引

和 英 索 引

～～～ **著 者 略 歴** ～～～

春日 健（かすが たけし）
福島県立会津高等学校 卒業
山形大学工学部電子工学科 卒業
山形大学大学院工学研究科修士課程電気
　工学専攻修了
現在，福島工業高等専門学校 名誉教授
博士（工学） 東北大学

濱﨑 真一（はまざき しんいち）
福島工業高等専門学校
　電気電子システム工学科 准教授
長岡技術科学大学 客員准教授
博士（理工学） いわき明星大学
　　　　　　　　（現：医療創生大学）

ⒸTakeshi Kasuga・Shinichi Hamazaki　2021

スッキリ！がってん！ 電気英語の本

2021年12月 3日　　第1版第1刷発行

著　者	春日　　　健
	濱﨑　真一

発行者　田　中　　　聡

発 行 所
株式会社 電 気 書 院
ホームページ　www.denkishoin.co.jp
（振替口座　00190-5-18837）
〒101-0051　東京都千代田区神田神保町1-3 ミヤタビル2F
電話(03)5259-9160／FAX(03)5259-9162

印刷　中央精版印刷株式会社
Printed in Japan／ISBN978-4-485-60049-8

• 落丁・乱丁の際は，送料弊社負担にてお取り替えいたします．

書籍の正誤について

万一，内容に誤りと思われる箇所がございましたら，以下の方法でご確認いただきますよう
お願いいたします.

なお，正誤のお問合せ以外の書籍の内容に関する解説や受験指導などは**行っておりません**.
このようなお問合せにつきましては，お答えいたしかねますので，予めご了承ください.

正誤表の確認方法

最新の正誤表は，弊社Webページに掲載しております.
「キーワード検索」などを用いて，書籍詳細ページをご
覧ください.

正誤表があるものに関しましては，書影の下の方に正誤
表をダウンロードできるリンクが表示されます. 表示さ
れないものに関しましては，正誤表がございません.

弊社Webページアドレス
https://www.denkishoin.co.jp/

正誤のお問合せ方法

正誤表がない場合，あるいは当該箇所が掲載されていない場合は，書名，版刷，発行年月
日，お客様のお名前，ご連絡先を明記の上，具体的な記載場所とお問合せの内容を添えて，
下記のいずれかの方法でお問合せください.
回答まで，時間がかかる場合もございますので，予めご了承ください.

郵便で
問い合わせる

郵送先

〒101-0051
東京都千代田区神田神保町1-3
ミヤタビル2F
㈱電気書院　出版部　正誤問合せ係

FAXで
問い合わせる

ファクス番号　**03-5259-9162**

ネットで
問い合わせる

弊社Webページ右上の「**お問い合わせ**」から
https://www.denkishoin.co.jp/

お電話でのお問合せは，承れません